The ISO 9000 Quality System

The ISO 9000 Quality System

Applications in Food and Technology

DEBBY L. NEWSLOW
D. L. Newslow & Associates, Inc.
Orlando, Florida

A JOHN WILEY & SONS, INC., PUBLICATION
New York · Chichester · Weinheim · Brisbane · Singapore · Toronto

This book is printed on acid-free paper. ⊗

Copyright © 2001 by John Wiley & Sons, Inc. All rights reserved.

Published simultaneously in Canada.

No part of this publication may be reproduced, stored in a retrieval system or transmitted in any form or by any means, electronic, mechanical, photocopying, recording, scanning or otherwise, except as permitted under Sections 107 or 108 of the 1976 United States Copyright Act, without either the prior written permission of the Publisher, or authorization through payment of the appropriate per-copy fee to the Copyright Clearance Center, 222 Rosewood Drive, Danvers, MA 01923, (978) 750-8400, fax (978) 750-4744. Requests to the Publisher for permission should be addressed to the Permissions Department, John Wiley & Sons, Inc., 605 Third Avenue, New York, NY 10158-0012, (212) 850-6011, fax (212) 850-6008, E-Mail: PERMREQ @ WILEY.COM.

For ordering and customer service, call 1-800-CALL-WILEY.

Library of Congress Cataloging-in-Publication Data:
Newslow, Debby L.
  The ISO 9000 quality system: applications in food and technology/Debby L. Newslow.
    p. cm.
Includes bibliographical references and index.
  ISBN 0-471-36913-6 (acid-free paper)
  1. Food industry and trade—Quality control. 2. ISO 9000 Series Standards. I. Title.
TP372.5.N48 2001
664′.0068′5—dc21   00-054573

Transferred to digital printing 2006

10 9 8 7 6 5 4 3 2 1

In Memory of Lilee Beary

CONTENTS

Preface — ix

Acknowledgments — xii

1 **Introduction** — 1

2 **The Certification Effort** — 14

3 **ISO 9001:2000—The Revision** — 27

4 **Getting Started** — 33

5 **Quality Management System (ISO 9001:2000 Section 4.0)** — 53

    5.1 An Overview, 53
    5.2 Quality System, 54

6 **Management Responsibility (ISO 9001:2000 Section 5.0)** — 62

    6.1 An Overview, 62
    6.2 Management Responsibility, 65
    6.3 Document and Data Control, 73
    6.4 Control of Quality Records, 82

7 **Resource Management (ISO 9001:2000 Section 6.0)** — 87

    7.1 An Overview, 87
    7.2 Training, 88

7.3 Process Control, 98
7.4 Handling, Storage, Packaging, Preservation, and Delivery, 102

8 Product Realization (ISO 9001:2000 Section 7.0)   108

8.1 An Overview, 108
8.2 Contract Review, 109
8.3 Purchasing, 112
8.4 Customer-Supplied Product, 119
8.5 Inspection and Test Status, 121
8.6 Product Identification and Traceability, 122
8.7 Control of Nonconforming Product, 125
8.8 Design Control (Design and Development), 127
8.9 Servicing, 132

9 Measurement and Analysis (ISO 9001:2000 Section 8.0)   134

9.1 An Overview, 134
9.2 Inspection and Testing, 135
9.3 Control of Inspection, Measuring, and Test Equipment, 138
9.4 Corrective and Preventive Action, 148
9.5 Internal Quality Audits, 164
9.6 Statistical Techniques, 182

10 Certification   184

11 Auditors Are Human   191

12 ISO 9000: The "Envelope" for the Food Industry (HACCP, GMPs, TQM, Malcolm Baldridge, and more)   197

13 Common Questions, Concerns, and Pitfalls   215

14 Summary and Conclusions   233

References   236

Index   238

PREFACE

Is there a place for ISO 9001 in the food industry? There is no doubt! Absolutely, there is a place for ISO 9001 in all food manufacturing and related fields. This statement is based on many years of firsthand experience with more than 100 different companies. As an industry professional, an assessor, and a consultant, I have seen the struggles and benefits from many different viewpoints, having been involved with many companies in various stages of registration. Some were in the very early stages. I had the opportunity of watching these companies progress from preliminary assessments to certification assessment. Other companies were registered long before I became involved with them. I saw the results through their routine six-month surveillance visits and three-year recertification audits. Every new experience reinforces my belief in the value of ISO 9001 in the food industry.

Personal experience has shown me that adhering to the ISO 9000 requirements provides the structure and discipline for a strong and efficient quality management system. The mature system prioritizes the foundation for continuous improvements, improved quality performance, and increased profits while meeting the customer's needs and expectations. ISO 9000 makes good business sense. A good friend of mine, Jon Porter, a consultant specializing in food safety, sums it up perfectly by saying that "ISO is the envelope, and everything fits inside."

It is frequently asked: "How difficult is it to achieve registration?" The answer is not simple. Each system is different. It begins with identifying and documenting existing processes, evaluating how each system meets the requirements of the ISO standard, and then addressing these gaps. Do not reengineer the process to meet the ISO requirements but integrate it into the

ISO structure. The road to ISO 9001 compliance is not candy coated, but it is worth the effort. Russ Marchiando, quality systems coordinator for Wixon Fontarome in Milwaukee, Wisconsin, described the certification process with the statement that "ISO is a journey not a destination."

WHAT IS THE REAL ISO EXPERIENCE?

What is the real "ISO experience"? An informal project was conducted prior to preparing this text. A cross section of individuals representing many different manufacturing and servicing organizations in various stages of ISO compliance were informally presented with a series of questions. These organizations related either directly or indirectly to the food industry. Participation was voluntary. Some associates asked to remain anonymous because either company policy did not allow them to be officially quoted or just because of the nature of his or her comment. Where appropriate, "thoughts" are integrated into the specific chapter topics. The following is a sample of the questions asked:

- What is the greatest/most useful process/system benefit experienced as a result of certification?
- What was the biggest surprise? Disappointment?
- Was getting certified worth the effort?
- Do you feel that you have received a return on your investment? How can this be measured?
- If you could do it all over again with what you know now, what would you do differently?
- Can you sum up in one or two phrases what compliance has really meant to your process?
- What was the hardest part of the system implementation effort?
- What has been the most difficult aspect of maintaining the system after certification?

This book is meant to serve as a technical reference, providing guidance and assistance in applying the requirements of ISO 9001 to food and food-related industries. Examples, ideas, and recommendations for the application of those requirements to different process situations is based on industry experience and expertise. The text addresses the concept, concerns, and potential pitfalls toward certification along with specific requirements and various options that can be applied. Frequently identified nonconformances, common questions, and the integration of existing programs such as Hazard Analysis Critical Control Points (HACCP) as related to ISO 9001 compliance will be addressed.

As will be discussed, the ISO 9000 series of standards has recently been revised into one compliance standard ISO 9001:2000. Comments and recommendations related to this revision are contained within this text, and references are also made to some specific wording of the ISO 9000:1994 series. This text is meant to provide information and guidelines to enhance the establishment of an effective quality management system compliant with the requirements of ISO 9001:2000. References to the wording of the 1994 version are included to enhance understanding and interpretation. The 2000 revision did include some rewording to provide a more generic basis for applying the standard to many different types of organizations. Our purpose is to enhance and provide the most effective understanding and interpretation as related to food and food-related operations. The reader should keep in mind that experience has been based on many years of exposure to both ISO 9000:1987 and ISO 9000:1994 series of standards. Although wherever possible, references and discussions as appropriate to ISO 9001:2000 will be included, the key at this point in time is to use the knowledge learned and shared in this text to provide the foundation for a strong and effective quality management system. One cannot teach experience! We will all grow and become "experienced" as we practice our "lessons learned." Compliance to the ISO standard does not guarantee quality, but it does provide a structure and discipline that results in a positive framework for the consistent production of a quality product.

DEBBY L. NEWSLOW
Orlando, Florida

ACKNOWLEDGMENTS

The content of this book is a culmination of many years of experience working and learning with many professionals in food and food-related industries. I would like to acknowledge and thank everyone that I have had the pleasure to meet and work with over the years. It is all this knowledge and wisdom from so many individuals that provides the foundation and wisdom for this text.

I would like to thank the following friends and contributors who specifically took the time to share their thoughts and experiences. The input from those who have experienced the experience will be valuable insight to every reader. A special thanks to the following: Charlie Stecher, Sue Goode, Jon Porter, Yvette Castell, Ed Steven, Victor V. Margiotta, Mike Burness, David Largey, Patti Smith, Jim Murphy, Tim Sonntag, Tom Marchisello, Al Gossmann, Keith Gasser, Andy Fowler, Gail Cartwright, Bill Lockwood, Naresh Modhera, Eric Halvorsen, Rick Aldi, Charlotte Sladek, Dave Demone, Mark Atkins, Brian Pugliese, Sylvia Garcia, Rick Bay, Dana Crowley, Russ Marchiando, Henry Gibson, Peter Gottsacker, Darren Weber, Ginna Young, Linda Taylor, Rex. N. Gadsby, Glenmore Wint, Karen Morgart, Jim Blaha, Candida Burgos, Bill DuBose, Brian Dunning, Linda Zastrow.

Thank you also to those who choose to remain anonymous. I would also like to thank those who spent countless hours proofreading the manuscript with total dedication and diligence: Sylvia Redditt, Martha Newslow, Nancy McDonald, Russ Marchiando.

ACKNOWLEDGMENTS

I would also like to thank Heather Newslow, the most efficient and accurate word process person, without whom I never would have gotten all the drafts completed and material submitted. And finally, I would like to thank my parents for all the years of encouragement and believing; and to all my family and friends for their encouragement and patience during this project.

1
INTRODUCTION

ISO 9000 is an international standard directed at the quality management process of an organization. Prior to ISO 9001:2000 the term ISO 9000 referred to a group of standards that included ISO 9001, ISO 9002, and ISO 9003.

An informal request was presented to a sampling of food and food-related industry professionals who had experienced ISO 9000 Quality Management activities. Quotations from these individuals are included throughout this text. A typical introductory question was "Why ISO 9000?"

Why ISO 9000?

> [Certification] promotes a system that anticipates and prevents problems rather than one that simply reacts . . . building quality in rather than inspecting it in.
> —Mike Burness, Director of Quality Assurance, Pepperidge Farms, Inc.

> ISO 9001 creates the platform by which a company can continuously evaluate and improve the business. Without ISO a company in the food industry may not have all the tools they need to ensure proper resolution of business opportunities. ISO 9001 implementation is no magic bullet. Many businesses have implemented ISO 9001 quality systems that merely comply with the standard and miss out on what is necessary for the growth and management of the business. I have seen ISO 9001 certified systems that do not meet proper regulatory and good manufacturing requirements to manufacture food products. Development of ISO 9001 needs to be a sincere effort by a properly trained management team in order to be a successful endeavor.
> —David Largey, Quality Assurance Manager, Campbell Soup Company

The answer to the questions "why ISO" and "did we receive a return on our investment" is very important. A return on investment can be defined in two primary ways. The first is based on a subjective comparison of where we were 3 years ago prior to our commitment to an ISO quality system and where we are now 18 months after certification.

Subjectively, I feel we have obtained adequate return on investment for the age of our system. Since the number of ISO certified companies in the U.S. Food Industry is small but growing, we are in select company. I feel it has helped us to secure new and existing business, because from a quality system standpoint, the elements making up the ISO standards are exactly what many of the major food companies in the U.S. and abroad seek (along with effective food safety systems like HACCP and GMPs and thorough sanitation programs). Before our ISO commitment our quality system was much more departmentalized and was primarily a quality control department responsibility. The certification process helped us to implement a lot of programs that were desired by the Quality Control Department years earlier but were never implemented due to lack of organizational commitment. Quality is now being communicated throughout the organization and there is a good framework in place to bring forth company-wide improvement.
—Tim Sonntag, VP Quality Assurance & Technical, Wixon Fontarome

"ISO" is not an acronym, but a term that means "equal" in Greek. The International Organization for Standardization located in Geneva, Switzerland, was founded in 1946 and is made up of approximately 100 countries, including the United States. Technical Committee 176 is a part of this organization and is responsible for defining and maintaining the ISO standards. The need for consistency and harmonization of international trade was developed through various requirements and criteria dating as early as the NATO documents AQAP-1 published in 1968. In 1983, as a result of trade interfaces and growing international technical relationships, the International Organization for Standardization established Technical Committee 176 to develop an international standard that focuses on quality assurance and management. This work, completed with the publication of the ISO 9000 series in 1987, was accepted worldwide through its adoption as the European Standard EN 29000:1987. Since its development, the ISO standard has been adopted as a national standard by more than 100 countries. The ISO 9000 series was revised and reissued in 1994 to further clarify and define its requirements. In 2000, the ISO standard was issued in a completely different format, focusing on the system's usefulness and effective maintenance of compliance status.

The ISO 9001 standard is a management tool that focuses on meeting the customer's needs and expectations. When integrated into a process, an ISO-compliant system provides the foundation and structure through documentation and objective evidence that promotes consistency throughout the entire operation. This formalized documented system clearly defines management policy, objectives, and expected performance. Compliance does not guarantee product quality but does provide the basis for a sound quality system. Continuous improvement is inherent in the system especially experienced through executive management review meetings, the corrective/preventive action, and internal quality audit activities (1).

"Quality management" refers to "all activities of the overall management function that determine the quality policy, objectives and responsibilities... within the quality system" (ISO 9001:1994). The term "quality system" is defined as "organizational structure, procedures, processes and resources needed to implement quality management" (ISO 9001:1994). The standard focuses on the existence, implementation, and effectiveness of the quality system, not the individual product. Certification to the standard means that the organization has a quality management system that meets the scope of the stated standard. It does not certify the product!

For the most part, the terms "certification" and "registration" are used interchangeably in the world of ISO and more specifically within this text; however, to enhance clarification, consider the following distinction as defined by Robert Peach (1997) in *ISO 9000 Handbook, Third Edition:*

> Certification [is the] procedure by which a third party gives written assurance that a product, process, or service conforms to specific requirements [whereas] registration [is the] procedure by which a body indicates relevant characteristics of a product, process, or service, or particulars of a body or person, and that includes or registers the product, process or service in an appropriate publicly available list (p. 19)

The ISO standard is purposely generic in nature and can be applied to a variety of services whether building a house, manufacturing TV sets, building airplanes, manufacturing chilled orange juice, making hamburgers, calibrating scales, or almost any other product or service being performed. Although the standard can be implemented by organizations, regardless of the type of product or service, it must be accurately interpreted and applied to the specific process. The standard must be complied with, but each process is unique. What works for one process may not be appropriate for another. The management team must define its system in a compliant manner useful to its overall process and business system needs.

As consumers, we return to our favorite products because we are satisfied. Adherence to the requirements of the ISO standard provides the basis for a sound quality system. Do not reinvent the process. ISO compliance is not a replacement! Existing HACCP (Hazard Analysis Critical Control Point), GMP (Good Manufacturing Practices), sanitation, pest control, and food hygiene programs should be integrated into the structured ISO system. This promotes adherence to defined procedures while enhancing the program through continuous improvement. HACCP and ISO together make good sense. The HACCP plan focuses on product safety whereas ISO focuses on the quality management system. ISO can provide the structure and discipline by which the HACCP plan can be incorporated.

The first ISO 9000 registration in the United States was in 1991. As of the year 2000 there were approximately 40,000 registered companies in the United States and Canada. Although food manufacturing companies have been approved in Europe and the United Kingdom since the initial acceptance of the standards, the first food companies approved in North America were the Lipton Company (Canada) and the M&M Mars Company (Waco, Texas) both in the first quarter of 1994. By the end of 1994, there were approximately 15 approved companies in the SIC (Standard Industry Code) category 2000 and by the end of 1995 approximately 35. Despite this slow start, there continues to be increasing interest. Approximately 250 related food manufacturing and service operations were certified within the food industry SIC at the time of this printing.

A current list of ISO 9000-approved companies by their Standard Industry Code can be acquired in the *ISO 9000 Registered Company Directory* available from McGraw-Hill, Columbus, Ohio. The prefix 2000 is for food and food-related companies. The scope of certification may represent a specific process, a specific location, or even several sites that perform the same operation.

Food industry professionals have been very skeptical of the benefits of an ISO 9000 compliance system. Generally speaking, food processors have stated that they feel that in meeting basic manufacturing standards and government requirements their processes are already adequately documented. Frequently, statements are made that the ISO standard does not apply to the food industry and that the requirements are too restrictive. However, as more and more food-related companies become familiar with ISO, it is being realized that compliance brings a structured quality management system to the existing family of quality and safety programs. Compliance provides the foundation for a strong, effective, and disciplined system that promotes customer satisfaction and continuous improvement in quality. Some companies make the decision to comply because it is virtually necessary for international trade. However, the majority do it because it makes good business sense.

Thoughts on Logic and Good Business Sense

> Perhaps the biggest surprise is that the standard requires only what makes good business sense. OUR RULE: If it doesn't make good sense we stop, reevaluate our interpretation of the standard and consider our alternatives.
> —Sue Goode, ISO Coordinator, Cargill Corn Milling

> I was pleasantly surprised at exactly how logical the ISO system is. I believe ISO should be viewed in the same context as GMPs or GLPs. ISO is simply Good Business Practices.
> —Jim Murphy, Manager of Design Process and Validation, The Dannon Company

ISO compliance set the "baseline" for other programs such as continuous improvement.
> —Rick Aldi, Director, Quality & Environmental Affairs,
> Hiram Walker & Sons, Ltd.

Benefits from Achieving ISO Certification

ISO 9001 certification benefits may be measured as "external" and "internal" benefits. External benefits are realized externally from the system such as an increase in market share or gaining customer recognition. An internal benefit is a benefit experienced within the system such as increased employee moral or a reduction in nonconforming product.

For those companies providing products to the European market, it is very likely that ISO certification is required in order to be considered as a supplier. This would be considered an external benefit. Realizing external benefits depend heavily on the specific business and focused market.

However, creating the environment for internal benefits makes good business sense when applied to any process. ISO 9001 certification does not guarantee quality, but it does provide a structure and discipline that result in a positive framework for the consistent production of a quality product.

A major benefit may identify activities within the process that have existed for some time and that no one really knows why. Sometimes the activity was the result of a memo or a suggestion implemented years past. Analysis of these activities, revising, or just discontinuing them can be a very positive cost-savings event.

> Several companies have told me that they were pleasantly surprised at the amount of activities in existence that have been being done for some time with limited value but significant cost to the process. Once these activities were identified, they were either revised into a useful function or eliminated all together.
> —Jon Porter, President, J. Porter and Associates, Ltd.

Further examples of some stated and observed internal benefits are:

- Improved efficiency through better documentation and communication
- Improved consistency of manufactured items
- Reduction in amount of rework and nonconforming product
- More efficient disposition and decrease in resolution time of hold and rework product
- Improved customer satisfaction
- Improved motivation and employee involvement through all levels of the process
- Reduced customer complaints

- Reduced customer audits
- Identified "nonvalue" and/or out-dated activities
- Defined and documented requirements
- Increased level of ownership from associates at all levels of the operation
- More structured associate communication process

System Becomes "a way of doing business"

> It was a pleasant surprise to me that such a diverse food ingredient company as ours could successfully meet a very aggressive 18 month timeline from when we decided to seek ISO 9001 certification to when we obtained it. However, it was an ISO 9001 quality system still primarily driven by the quality department. The biggest surprise was how our system has grown in the 18 months since certification to become more of a "way of doing business" management system. Once executive management became more involved in securing and communicating this as "our quality system" to all levels of the company it became a true quality improvement system for the entire organization. In fact, all members of executive and upper level management are now intimately involved in our system as both internal auditors and as attendees at our quarterly ISO management review meetings.
> —Tim Sonntag, VP Quality Assurance & Technical, Wixon Fontarome

What Were the Greatest/Most Useful Process/System Benefits Experienced as a Result of Certification?

An anonymous contributor provided the comment that "ISO provides us the discipline to follow through."

Rick Aldi, Director, Quality & Environmental Affairs, Hiram Walker & Sons, stated that the

> best thing for us regarding certification is that everything is in one place and... people know where to find it. It [certification] created a discipline. The benefits of the integrated systems is better visibility of requirements throughout the organization, better communication between operating and business units, lower inventories, better forecasts, and on and on.
>
> ISO certification has allowed us to better utilize the [ISO] standard as a tool to keep our business process on track and measure our forward progress.
> —Sue Goode, ISO Coordinator, Cargill Corn Milling

> As a service organization, our corporate quality auditing group maintains ISO 9001 certification. Operating a quality management system that is compliant with the ISO 9001 standard has driven us to focus on delivering work process and service outputs that are meaningful to our internal customers. As a result, our work processes have become much more efficient to the point where we have been able to meet our goals and objectives with 30% less "people-hours" [while]

continuing to reduce error and repeated tasks. We have also been able to identify and eliminate "non-value added activities."
—Tom Marchisello, Director of Quality Assurance, Campbell Soup Company

All establishments I've had the opportunity to work with have indicated the same thoughts, "We didn't know we were doing that." There has not been one manager or administrator who has not made the same declaration. The greatest benefit is simply getting to know what is going on in the plant and who is doing what [what is and is not being done].
—Jon Porter, President, J. Porter and Associates, Ltd.

I think the most useful benefit has been a heightened level of executional discipline resulting from system requirements as they relate to process documentation. We have always operated in an "experiential" industry and with the implementation of ISO 9002, many of the daily processes we have historically performed, even consistently, have now been documented and improved. This has helped us evolve many aspects of our quality system from experiential to process driven.
—Mike Burness, Director of Quality Assurance, Pepperidge Farm, Inc.

Reckitt & Colman's customers were not asking us to get certified, we wanted to do this for our benefit. The most useful process experienced as a result of certification was a written quality system that had plant wide acceptance.
—Charlie Stecher, Quality Assurance Manager, Reckitt & Colman, Inc.

Benefits that we have experienced include (1) the organizational focus (not just one department) on improving and expanding our quality system and (2) the clear definition of responsibilities. Teams know what to do. Procedures are documented.
—Henry Gibson, Quality Assurance Manager, Campbell Soup Company

One of the greatest benefits that we as a company have experienced was in the documentation of the quality system (work instructions, procedures, specifications, etc.) and the unity seen in the effort to be ready for certification.
—Yvette Castell, Quality Assurance Manager, Dairy Industries (Jamaica) Limited

The greatest benefit from implementing a quality system that meets the ISO standard is that the system and implementation force a discipline into the operation. It is a discipline to precisely follow procedures and specifications. This discipline becomes the foundation for continuous improvement efforts because, once processes are standardized and [followed by] everyone, people identify ideas that would improve the documented process. The improvements then become the standards, which everyone follows, until the next improvement is identified. This discipline helps ensure that the best practices are followed.
—Rick Bay, Plant Manager, Reckitt & Colman

The greatest benefit as a result of certification was the initial aspect of having to do an in depth analyses of the total business, and to document the systems and processes to the ISO standards. The biggest surprise as General Manager at the time was the willingness and dedication of all employees at every level to par-

ticipate in the certification process. If I had a disappointment it would be for not having done it sooner.
—Rex. N. Gadsby, retired GM and CEO Dairy Industries (Jamaica) Ltd. and Grace Food Processors, Ltd., Jamaica, West Indies

Bill Lockwood (Package Quality Manager, Hiram Walker & Sons, Ltd.) summarized some of their benefits:

One of our greatest benefits has been the increase in quality awareness. Employees now tell us to do it right the first time. [Employees] have pride [in being] certified [and in] being a quality place to work. Now when we have a potential problem, the employees will state "we cannot do it that way because it is not [defined in the] ISO [procedures]." What they really are saying is, "This is not quality, this is not following procedure." Employees have bought into this QMS [Quality Management System].

Overall consistency—product, process, and business consistency. Thus better product quality and quality service to our customers.
—Ed Steven, Plant Manager & Patti Smith, Quality Assurance Manager, NZMP

ISO 9001 did not directly improve our business profitability, however we both manage our business and communicate more effectively as a result of ISO 9001. The business is recognized as a well managed company and we have received new business as a result.
—Dave Largey, Quality Assurance Manager, Campbell Soup Company

As a result of certification employees are now very conscious of adhering to proper procedures in their respective areas. This leads to a consistent level of quality. There has been a remarkable improvement in inter-departmental communication at all levels.
—Andy Fowler, Research Engineer/Management Representative, Bacardi & Company Limited

The compliant system provides formality and structure. We have set an internal system standard that we are required to maintain.
—Gail Cartwright, Assistant to the AVP Human Resources Department, Bacardi & Company Limited

Numerous management changes due to an acquisition required a solid mature system to help our organization through this time. Because our quality management system was founded more on the philosophy of certification rather than business improvement, we struggled when the business environment became more competitive. As one that is involved in overall business improvement, I am amazed at the ability of the ISO reviews to measure with reasonable accuracy the state of our business. When our quality management system was struggling, there were numerous other areas of our business that were struggling.
—Anonymous

The following summarize further thoughts on what were the greatest, most useful process/system benefit experienced as a result of certification:

INTRODUCTION

- "The most useful benefit from registration is the ability to define and communicate what our quality system is." (Al Gossmann, Quality Assurance Manager, Cultor Food Science)
- "Increase team work [such as] audit teams, CAR [corrective action] teams, rework reduction teams." Anonymous
- "Increased employee involvement [through the] teams, suggestions, writing work instructions [etc]." Anonymous
- "Consistency of procedures [knowing that] this is a best practice. Improvement in procedures [has] eliminated unnecessary steps." Anonymous
- "Improvement of processes [has been experienced through] corrective actions, audits, [and] . . . receiving [activities]." Anonymous
- "New procedures [were developed for handling] quarantine [hold product], calibrations, [and] document control." Anonymous

Many state that if they had the opportunity to do it all over again they would not be so aggressive with their timeline. Depending on the size and process variations, on the average, it takes a company 18 to 24 months to complete the initial phase. The initial phase of certification includes defining, implementing, controlling, and maintaining the process in such a manner that it provides significant objective evidence of compliance. The process is not simple and requires a strong commitment from management for both hands-on involvement and for providing adequate resources to the system. It is more important to take the time to define the system, manage it through training and internal audits, and then effectively address existing and potential noncompliances through the corrective and preventive action process. A well-defined management review process should be applied to monitor the system for effectiveness and suitability applying resources as appropriate.

Once the system is initially implemented, approximately 3 to 6 months of objective evidence (records) demonstrating compliance to defined requirements is usually required before the system can actually become certified. The registrar will evaluate the system as it is defined and in doing so must see evidence that it not only has been defined but that it is actually being maintained compliant to the requirements of ISO 9001. Evidence in the world of ISO 9001 means records. Quality records must be maintained to demonstrate compliance to specified requirements.

It Is Not Magic That Makes It Happen

It is not magic that makes it happen, but the diligence of all of your departments uniting on a common process, performing your regular jobs in addition to implementing your ISO system. It will bring you closer together as a company with better communication. It is important to be sure that ownership of the quality

system belongs to the whole staff. Don't maintain your quality system just to satisfy your assessor. Your system should always focus on keeping your customers happy and identifying improvement opportunities.
—Charlotte Sladek, Business Development Representative, Lloyd's Register Quality Assurance (LRQA)

We strive for excellence not perfection within our system. No system is perfect. The system as one unit with total people involvement from all levels of the operation is a major advantage of an ISO compliant system.
—Karen Morgart, Packaging Specialist, Hiram Walker & Sons, Ltd.

Getting Started

Getting started can be tough. The next chapter is devoted to the decisions and information needed to begin the implementation process. Management commitment and team involvement are inherent throughout the whole process.

One of the hardest parts of the implementation effort was getting the employees and managers to buy into it. Convincing them that we were serious about developing, implementing, and maintaining the system. This was not going to be the flavor of the month program.
—Dave Demone, Environmental and Quality Control Manager, Domino Sugar

Getting associates to understand and fully embrace the system was one of the hardest parts of the implementation effort.
—Andy Fowler, Research Engineer/Management Representative, Bacardi & Company Limited

Thoughts on Marketing ISO 9000 Compliance Within Your Organization

Mark Atkins, President of ISO Quest Consulting Services, Inc., St. Petersburg, Florida, has had considerable firsthand experience marketing compliance within an organization, as a consultant and as a certification auditor.

Now that nearly 40,000 companies in the United States have been registered to ISO 9000, it has become almost cliché that strong management commitment is an absolute necessity. I do not argue with this. Failure is inevitable without management's understanding of the scope of the work and the resources required to ensure the quality system is effectively implemented and maintained. Simply having management's buy-in is not enough. Commitment and involvement is needed at all levels, not just at the top.

Perhaps the best-documented system I've seen as a consultant and auditor over the last few years was one that generated the least amount of enthusiasm. The rank-and-file employees were stumbling over their procedures and saw very little use for the program. ISO was to them a management "add-on." When I asked top management why the "energy" for ISO was so low, they said employees were not "using" the system. With a little investigation, it was obvious why they weren't "using" the program—it was never theirs to begin with. The ISO documentation mostly came from a "sister" plant and few of the employees were asked to review this documentation and fewer were made aware of what ISO 9000 could do for them. None were asked to be an internal auditor outside of Quality Assurance employees.

Occasionally I see the enthusiasm and energy coming from the bottom of the organization. Employees are tired of relying on "tribal knowledge" and they need good reliable information for doing their job and training others. They grow weary of watching chronic problems fester with no mechanism for correcting them. They see ISO as a godsend. They try to light the fire so management will understand what an ISO program could do for them.

And then there is supervision. Caught between the demands of their superiors and balancing the workload of their subordinates, they sometimes see ISO as another program, as one more thing to add [to] their plate. Is it any wonder they resist? They've heard the ISO horror stories and see quickly that this is a very long and laborious program, one that probably will impact them the most in terms of change to practices. ISO is perceived as something that will somehow restrict the informal manner in which they manage their day-to-day madness.

So how does one avoid alienating all levels of the organization with an ISO implementation? [One must] sell the program. By all accounts, ISO 9000 requires a lot of work and commitment. It takes an organization on average about 18–24 months just to implement. It requires work at all levels and affects activities and interactions across various groups. It usually requires changes in the way [one] manages and works. Oftentimes, ISO 9000 invokes a cultural change.

How then [does one] pull the organization together to effectively implement ISO? This may scare those of us in the Quality arena, but ISO implementation requires a well-thought out and well-constructed marketing strategy. "Marketing strategy" entails consideration to the product, the market segments, the communication strategy, the delivery system, and the price.

Product

The first consideration is for the company to decide what they expect to gain from ISO 9000? What are the expected benefits? As Steven Covey puts it, "start with an end in mind." Of course, the desired benefits depend upon [the] unique situation and culture of the company, but they can gain some insight by asking other registered companies how ISO has helped them. Some of the more common benefits have been: a gain in competitive advantage in the market place; a means of "spring-cleaning" of procedures; a useful framework for managing business processes; a structured means for reducing waste and redundancies; and improved organizational communication. Management should remember not to dictate all of the expected benefits. A core team from all levels of the organization should be educated on ISO 9000 and then have discussions on how their group could benefit from a well-documented, quality system that is rooted in commonsense.

Segments

In marketing a product, it is critical to remember the target audience. One cannot expect to generate much enthusiasm for benefits unrelated or seemingly irrelevant to their group; in fact, selling such benefits could evoke great resistance. As Ralph Waldo Emerson stated, "nothing great was ever achieved without enthusiasm." Everyone wants to know "what's in it for me." Again, the ISO team should decide in the beginning how each group will benefit. It's not a matter of hiding the overall benefits, but a matter of stronger emphasis of the ones most relevant to that particular group.

Communication Strategy

Now that the company has a clearer understanding of the expected benefits for the various segments of the organization, they will need to consider how will ISO be brought to the user. Here are some tips on developing a sound communication strategy:

- Ensure employees that the head of the organization is fully supportive of the program.
- Appoint a cross-functional communications team.
- Dedicate money for communications, develop consistency in your message, gear presentations to your audience, extolling the expected benefits.
- Give the audience a feel for what's involved, use the media mix you have such as the company newsletter, magazine or communication meetings.

Most of all remember that timing is everything. For example, don't go around telling union employees that ISO is a good way to document how they do their job if a strike is possible.

Delivery System

The communication strategy will be the means of how the employee will be made aware of ISO. A delivery system, however, will be needed to induce employees to "use the system" and help persuade them to continue using it. Will the management representative be the primary means of delivering the product, i.e., guiding procedure development and training? Will the department manager or supervisor dictate which procedures or work instructions will be written or will they involve their employees for their input and involvement in writing their procedures and ensure proper execution? Remember the previous example where employees did not use the system. Sound guidance from the management representative, a qualified consultant, and their management is very helpful, but don't discount the necessity of employee input. They are the experts in their activities and it is their ownership in the long haul will sustain the program.

Price

Any marketing strategy, of course, should give consideration to the price of the product relative cost compared to the benefits. It is not unusual for organizations that did not "calculate the cost" of the program initially to drop the program. Sources of these costs may be from initial training of staff, internal auditor training, consultant fees, gap analysis, reassessment and initial registration fees, employee overtime, and communication costs. There are also some costs in terms of employee time that may not affect the budget, but may come at the expense of other programs. Management will need to consider if the "price" of the ISO program outweighs the benefits. ISO is an investment and the true benefits may not be realized until after registration.

ISO 9000 can definitely improve the overall efficiency and effectiveness of the organization. Again, using a well-crafted marketing strategy at the onset will ensure a smoother implementation and enhance organizational focus.
—Mark Atkins, President, ISO Quest Consulting Services, Inc.

ENDNOTE

1. Clements, Richard Barrett. 1995. *Quality Manager's Complete Guide to ISO 9000; 1996 Cumulative Supplement.* Prentice Hall: Englewood Cliffs, NJ.

2
THE CERTIFICATION EFFORT

ISO implementation and its value depend on the management team and the specific operation to which it is applied. As will be evident throughout this text, defining and implementing a quality management system compliant with the requirements of ISO 9001 is no easy task.

No matter how easy it looks on TV, a professional baseball player will state in no uncertain terms that making it to the World Series is no easy task. The foundation for these achievements is generally related to every opportunity, accomplishment, and disappointment that one has experienced. Opportunities, accomplishments, and disappointments are terms that one can use to describe the implementation effort. There will be days when the walls and documents feel like they are closing in. Then the very next day there will be opportunities and positive advances that make you wonder how you ever lived without them. This may sound like an overglorification. As an assessor, often I visited a facility that had made it through the maturity process. Having been certified for a year or more, together we would remember the preliminary assessments and the expressions of discouragement. It is a great feeling to see smiles when I ask employees if they ever thought they would be where they were now. The smiles and expressions of accomplishment are so much more dynamic than those of gloom.

Compare the certification effort to building a house. Standing on the empty lot, full of trees and weeds and wondering how it will all get done. But the day comes and it does get done. And when it is done, it will provide a safe and enjoyable haven. I remember my first week of college. I had more books to read that first term then I had read my entire senior year of high school. I wondered how I would ever make it. Now I look back and think that it wasn't so bad after all and I am so thankful that I never gave up. Defining and implementing the ISO-compliant system is equal to all of these scenarios. It is not easy. It takes focus and team effort with everyone applying their expertise and resources. Everyone means everyone throughout all levels of the operation. Executive management must be involved. Executives must provide the opportunity for associates to be involved. There are some great and not so great days of accomplishment. It takes time. It takes commitment. Standing on that vacant lot, one could decide that a house isn't worth the effort and pitch a tent. It would be easier, it would take less effort and time. But in the end, one still only has a tent.

> It's like "Climbing the Blue Mountain" and every bit of the same great experience once achieved. In our business (restaurant and hotel) we have defined "quality" as being about "System Dependent and Personality Delivered."
> —Linda Taylor, Training and Quality Manager, Le Meridien Jamaica Pegasus

> The most useful benefit of registration is the ability to define and communicate what our quality system is. Many of the elements were already in place, but we could no longer reference our previous sister company.
> —Al Gossmann, Quality Assurance Manager, Cultor Food Science

It is very important to the certification effort that realistic and achievable goals be identified. Tim Sonntag (VP Quality Assurance & Technical, Wixon Fontarome) had an interesting comment regarding the initial certification effort:

> If we did this over, we would set a more realistic timeline based on the extent of each manager's level of commitment and "their" understanding of the changes "they" need to implement to make "their" departments ISO compliant. QA would be a resource for them to use to help keep their system ISO-compliant but each manager would develop their own systems and the documentation to go with it. I would also try to have more in depth executive management training on implementing an ISO 9001 quality system so that their knowledge of the process would be clearer. Finally, I would set up a semi-monthly meeting with the executive management team to discuss the status of the implementation process and any issues that required their attention or their action. We found that communication to the executive management level is crucial to the successful implementation of a quality system that is developed by managers and supervisors from all levels of the company.

A common theme among many of the respondents to our survey emphasized the resulting teamwork and bond that was felt as the system was being implemented. Many also stated that this bond has continued after certification was obtained. These types of comments are so prevalent among associates throughout the industry. Some examples follow:

Certification resulted in a bond among employees and departments (and even various sites to some degree) and a common objective to a far greater degree than any other initiative experienced during my twenty plus years career. It was accepted as much more than the flavor of the month promoting the majority of the practices to become ingrained and part of the culture. The system gave the operators the empowerment to do the job RIGHT, without being overruled by their foreman, supervisors and managers. It gave the operators the sense that they were an integral part of the process and that they could have a say in how operations were performed. This was a real morale booster!
—Dave Demone, Environmental and Quality Control Manager, Domino Sugar

Our success was directly related to dedication and efforts of our team. We believed in our team, we wanted them to be recognized in the global market place. Accomplishing certification brought it all together.
—Linda Taylor, Training and Quality Manager, Le Meridien Jamaica Pegasus

Developing & implementing a good quality system based on ISO 9001 and GMP principles will enhance a company's growth and maturity by maintaining the company culture and good foundation. This may require discipline, as some people don't like change and resist it. The greatest benefit of an ISO system is the linkages between internal audit findings, consumer/customer complaints, management review and corrective/preventive action processes. These by design promote ongoing continuous improvement. It is not difficult to maintain ISO registered status as long as there is management support and personnel who follow the procedures (process control).
—Naresh Modhera, Manager, Quality System, Reckitt & Colman, Inc.

When the question, Was getting certified worth the effort? was posed, varying opinions were received, many of which centered on management and associate involvement and the resulting team's success. This led to an effective and efficient quality management system that grew into a strong foundation for continuous improvement. As systems matured, many stated that executive management support continued to grow. Sales and marketing felt that the investment was paying off especially for those companies that were involved with ingredient manufacturing. Others commented that it was worth the effort because of an increase in the effectiveness of training. This was credited to the specific training criteria defined for becoming "qualified."

Tom Marchisello, Director of Quality Assurance, Campbell Soup Company, stated that "[as] our quality system is beginning to mature, . . . upper management has assumed an increasing role of responsibility in this system." Marchisello further stated:

As a corporate function we are expected to lead by example. Our certification is a very effective tool for leading the way for our manufacturing plants since we expect them to all establish, implement, and maintain a quality management system based on ISO 9001. Our return on the investment has been significant, especially when defined in terms of reduced waste and inefficiencies.

One common theme stressed throughout this text is the fact that compliance with the requirements promotes continuous improvement. Some may argue that this is not true, but in these instances I would challenge the effectiveness of their management review, corrective/preventive action, and internal audit processes. Many professionals have contributed thoughts on just this aspect. Following are a few examples:

> Certification was certainly worth the effort. As our company continues to grow, certification has provided us a springboard from which to improve. As standardized quality system requirements become more accepted around the world, we are in a better position to respond to market and customer demands for established and recognized quality systems.
> —Mike Burness, Director of Quality Assurance, Pepperidge Farm Inc.

> Without a doubt it has been well worth the effort. It has given us the structured common framework throughout our organization to move forward with continuous improvement efforts. In becoming certified, we communicate more effectively and are much more consistent in all aspects of our operation.
> —Keith Gasser, Quality Systems Manager, Tropicana

> Getting certified was indeed worth the effort. We have processes in place to help us grow and to be proactive and not constantly putting out fires. We have documentation to verify certain activities have been accomplished, record findings for future reference, and define our improvement cycles. All of that helps us to manage business more efficiently.
> —Sue Goode, ISO Coordinator, Cargill Corn Milling

Management Must Maintain Its Focus on Continuous Improvement!

> Our initial effort in '94 was necessary to compete in European markets. However, it became more of a status symbol then a method of business improvement. Unfortunately that is not a positive motivation to drive continuous improvement and improved customer satisfaction through the systematic implementation of a QMS. It is very important that the system focus is driven through continuous improvement and customer satisfaction.
> —Anonymous

Certification Made "System Activities so Much Clearer" and "Instilled Rigor"

> Yes it was worth the effort. It brought our team efforts to new highs making requirements and system activities so much clearer for everyone of our hotel associates.
> —Linda Taylor, Training and Quality Manager, Le Meridien Jamaica Pegasus

> The journey toward certification has resulted in significant improvements and instilled more rigor into how we run our business.
> —Henry Gibson, Quality Assurance Manager, Campbell Soup Company

Certification Enhanced Accountability and Team Involvement Creating an Overall Accomplishment of which to Be Proud

> Getting certified was worth the effort (although I might not have said so at the time) because it is an outside/independent validation of the system. It is where the rubber meets the road. (For example, you must do more than talk a good game.) The other value of the certification is that there is a tremendous amount of learning that takes place for everyone involved. The third reason, and maybe most important is that getting certified is something to be proud of, and everyone in the facility feels a part of this accomplishment.
> —Rick Bay, Plant Manager, Reckitt & Colman, Inc.

> One of the biggest surprises has been the overall unified pride all employees have taken in becoming certified. Everyone was very excited and we held a big celebration. It has carried over into an increased pride in our company as well.
> —Keith Gasser, Quality Systems Manager, Tropicana

> I personally think that [certification] was very worthwhile. It is quite an accomplishment for the plant and something to be very proud of. Perhaps the most important factor is the sense of discipline that has taken root in order to maintain certification. While I wouldn't consider the road to establishing a quality management system easy, it would be very easy to put all the documents in a corner and only pull them out to show customers when they want to see what you do. Maintaining a certification forces us to use and build upon our system. This takes tremendous discipline and that is something we needed. Furthermore, life is so much easier and stress free. While being a bit on the lighter side, one thing that has been a plus is customer questionnaires. Being able to check off the box in customer questionnaires that says: "If You Are ISO 9000 Certified, You Do Not Need To Fill Out This Questionnaire," has made the whole effort worthwhile.
> —Dana Crowley, Production Manager, Danisco Sweeteners, Inc.

> Getting certified was worth the effort, because it gives everyone a sense of accomplishment and a milestone to look back on and say "we did it." But the effects of the effort are far more valuable in the respects of having a written quality system.
> —Charlie Stecher, Quality Assurance Manager, Reckitt & Colman, Inc.

Boosted Customer Respect, Confidence, and Satisfaction

Andy Fowler (Research Engineer/Management Representative, Bacardi & Company Limited) stated that there was no doubt that certification was worth the effort:

> As the number one spirits manufacturer, our quality system has been certified to comply with the ISO 9002:1994 standard by a well respected registrar. This can only boost consumer respect and confidence in our product.

> Yes, certification was worth the effort. We experienced improved customer satisfaction, maintaining existing customers and gaining new customers.

Quality improvements resulted in a decrease in nonconforming product manufactured.
—Ed Steven (Plant Manager) and Patti Smith (Quality Assurance Manager),
NZMP

Is There a Return on Your Investment? How Can This Be Measured?

Answers varied to these questions, but the most frequent response was that effectiveness is demonstrated through a decrease in customer complaints and nonconforming product, including both in-process and finished product. One manager stated informally that he still had nonconforming product, but associates used the system's structure and definition to identify these issues while they were relatively minor. What might have been 5000 cases on hold in the past now may only be 10 or 15 cases. It is very difficult to measure what might have been without the system, but the confidence that the processes are in control was a common statement among many. Respondents felt that although it may be hard to measure, actual benefits to the overall operation and consistency of the process made it worth the investment. The system as it matures promotes continuous improvement throughout.

> Compliance has enhanced our system's consistency across the entire business and our ability to focus on process improvements.
> —Ed Steven (Plant Manager) and Patti Smith (Quality Assurance Manager),
> NZMP

Thoughts on the Return on Investment

Russ Marchiando (Quality System Coordinator, Wixon Fontarome) stated:

> From the quality assurance point of view, I believe that there has been a significant return on the investment. Many programs that were lacking have been instituted, lines of communication have been developed and a structured system for improvement has been established. These changes themselves are a significant return on the investment.

Yes, we have received a significant return on investment. Tangible returns include reduction in consumer complaints, improvement in product conformity, and reduction in nonconforming raw materials. Returns that are less clearly measured but may be more important overall include improved customer focus, clarified roles, responsibilities and accountabilities; and development of a continuous improvement ethic. Above all ISO has provided a solid foundation on which to build other programs such as HACCP.
—Jim Murphy, Manager of Design Process and Validation,
The Dannon Company

There should be a return on investment, and it should be measurable in reduced waste and rejects as well as improved customer service. After being certified

approximately one year we have not seen a step function improvement but feel we are close.
—Rick Bay, Plant Manager, Reckitt & Colman, Inc.

Getting certified was worth the effort, for two main reasons, one being the benefit of having employees more focused on following defined procedures and systems and secondly knowing that finished products now meet specification on a regular basis. The return on investment was quickly apparent and financially measured by rework/returns through doing it "right" the first time.
—Rex N. Gadsby, retired GM and CEO Dairy Industries (Jamaica), Ltd./Grace Food Processors, Ltd.

We have seen evidence of a return on our investment through the reduction in consumer complaints. We have also seen a reduction in the number of nonconforming intermediate and final packaged products.
—Andy Fowler, Research Engineer/Management Representative, Bacardi & Company Limited

We obviously have seen a return on our investment however, it will be quite awhile before we will recoup the total investment. As we have only been certified for nearly 6 months now, I cannot say that we have generated anymore business because of the certificate. I have had a few Sales & Marketing folks ask me for information regarding our certificate. Perhaps the easiest measurements we have are in the reductions in nonconforming product and reductions in the number of customer complaints. Dollars saved in reworking costs can be calculated very easily. Customer complaints, while easy to count and keep track of, are difficult to put a price tag on. The cost of lost opportunity associated with a customer complaint is an intangible that is very difficult to measure. At this point, the best measurement would have to be monitoring the number of complaints.
—Dana Crowley, Production Manager, Danisco Sweeteners, Inc.

As discussed in several of the comments so far in this section, many feel that one of the major indicators to measure return on investment is through the reduction of rework or nonconforming product. This is typically a very common statement heard from many professionals. Following are more thoughts on this aspect:

Return on our investment can be measured in the reduced number of rejects and reduced amount of rework, including the fact that overall we have had fewer "big" rejects.
—Charlie Stecher, Quality Assurance Manager, Reckitt & Colman, Inc.

Yes we have received a return on our investment. Our quality metrics (reduced consumer complaints and higher product conformance) have improved dramatically.
—Henry Gibson, Quality Assurance Manager, Campbell Soup Company

Yes, we have definitely received a return on our investment. This can be measured by looking at the reduced cost for nonconforming and rework products.
—Yvette Castell, Quality Assurance Manager, Dairy Industries (Jamaica) Limited

Keep in Mind That Basically You Get as Much [or as Little] as You Put in to [It]

I believe that in part you get as much out (return) as you put in to the certification. The other part, is how well the system is designed to fit within the business strategy.

—Anonymous

Defining the Measurable Business Effort

We feel that we certainly have received a return on our investment. Most of the investment was obviously in the form of resources, mainly for system documentation and development. Once that part was complete, it was actually fairly easy to implement the system, albeit a long process. Measuring the return on investment has been a more difficult task from a management perspective. Intangible measures include more disciplined processes and outputs, better-trained employees, etc. As we continue to improve the system, we are now taking it to a new level. As the QA field evolves, we will need to become more fluent in the language of management (i.e., economics). If we cannot communicate the value of the system in management terms, we will struggle to provide measurable business benefits. The certified system provides a framework from which a complete economic quality model can be easily developed. This includes measures related to the cost of nonconformance and direct measures for implementation of corrective actions. Listed below are a few examples:

1) *Cost of nonconformance; Increase in yield or losses prevented from incoming inspection and test.* Taking a look at incoming ingredients and detecting a nonconformance obviously saves valuable system costs. Downtime, poor product output, etc. are a few items that would be negatively affected by an ingredient non-conformance. Capturing this non-conformance prior to use sounds like QA 101, but there are many instances where it is not practiced. The system, coupled with the economic measures, provides a framework for assigning economic value to this task.

2) *Establishment of corrective action costs as they relate to return on investment.* As nonconformances are detected, implementation of *root cause* corrective action (typically time consuming and expensive), can be linked to an ROI (return on investment). Taking these measures into account now make the system a company investment instead of a cost, resulting in the ability to translate quality directly to the top and bottom lines, in economic terms (what a concept?).

In my opinion, if we, as a Quality Assurance function, do not begin the translation of the system into financial terms, we will not be able to evolve into business partners and will therefore fail to take advantage of system improvements as a measurable business benefit.

—Mike Burness, Director of Quality Assurance, Pepperidge Farm Inc.

Some companies feel that the overall process improvements make the entire investment worthwhile.

I talked with one company who did not care whether they received a return on the investment because the certification was considered a "cost of doing business." I have interpreted this to mean that certification is well worth the cost and that it is such a worthy pursuit that attempting to "cost-it-out" will separate the "system" from "production." "Production" is the tool for profit. Segregating work centers into cost centers is destructive and does not add to profit.
—Jon Porter, President, J. Porter and Associates, Ltd.

We have definitely received a solid return on our investment, however we have not quantified this measurement. There have been many accomplishments that have been associated with our ISO certification effort, each one of which individually would have given us an adequate ROI. It is really immaterial whether or not these accomplishments would have eventually occurred without the ISO 9000 initiative. The ISO certification effort has provided the common basis to make these accomplishments more easily attainable.
—Keith Gasser, Quality Systems Manager, Tropicana

Every phase of the entire ISO compliance effort can be related back to "management commitment." Mr. Sonntag emphasizes this aspect when discussing return on investment.

The most significant return on investment from a subjective standpoint, in my opinion, has been the significant role that upper management has taken in being responsible for the effectiveness of the quality system. Not only is every manager an internal auditor, but many of these same managers are on our Steering Committee, which is responsible for guiding our corrective and preventive action process. In addition, there is a weekly QA meeting with the President of our company to discuss quality system issues.
—Tim Sonntag, VP Quality Assurance & Technical, Wixon Fontarome

What Was the Hardest Part of the Implementation Effort?

When discussing what was the hardest part of the implementation effort, many agreed that it could be related to management and employee buy-in, commitment, and ownership.

Without a doubt the hardest part of the implementation effort was convincing the longstanding management employees of the need for change. Many of our employees have grown up with the company and defined their positions along the way. Along comes a new quality system implementation effort that requires them to place some parameters on their jobs and the way they operate their departments. This was a fundamental change for these people and a very difficult one to sell. It was also difficult initially getting management to support the effort beyond the meeting room. Certification was clearly a goal of the company's management. However getting management to "roll up their sleeves" and contribute towards the implementation effort was a difficult prospect. Everyone wanted to eat the vegetables from the garden but no one wanted to do the weeding or watering along the way. It took a long time to help management understand that the ISO 9001 system was a business system and not just a project of quality assurance. Once this was established at the very top of the organization, the system made a wholesale transformation and migrated towards a value added system. Finally the system began to work for us.

—Anonymous

The hardest part of the implementation effort was in getting a system that everyone bought into and felt ownership of, but also was effective and manageable. This included the decision of what to document, what not to document, and how to control all this documentation.

—Sue Goode, ISO Coordinator, Cargill Corn Milling

The hardest part of the system implementation is to get the understanding, buy-in, and commitment from the day-to-day users. The overall response is that "I know my job, why do I have to go through all this?" Afterward, some of the biggest detractors up front become the loudest spokespeople.

—Rick Bay, Plant Manager, Reckitt & Colman, Inc.

I think there were two difficult hurdles for effective and successful implementation. The first and most important is obtaining and maintaining executive support for the system. Implementing a system of this scope requires economic support, as well as consistent communication of the same message throughout the company. Although not a difficulty in our particular case, without it we would never have gotten out of the gate. The second difficulty is garnering support of the employees and keeping them abreast of the purpose/benefits of the system. This becomes particularly difficult when turnover occurs or a temporary workforce is employed. Keeping up with the employees, training them, and having them be a part of the system is a difficult proposition during implementation. Once mastered, however, it is the key to success.

—Mike Burness, Director of Quality Assurance, Pepperidge Farm Inc.

The hardest part of the system implementation effort was maintaining individual commitment to additional work, whilst going about business the old way.
—Rex N. Gadsby, retired GM and CEO Dairy Industries (Jamaica) Ltd./Grace Food Processors, Ltd.

More on Culture Change and Management Commitment

Clearly the hardest part of the implementation process was changing the culture of the company's employees, specifically the low and middle managers that had been with the organization for over 10 years. These managers had been part of the rapid growth experienced from a small regional food ingredient manufacturer to a much larger, more diverse co-packer and national supplier. Over the years, these managers defined their jobs and what was important for them became the way things were done. Each department had their own way of handling their operation and they had, in effect, built walls around their department that made many quality department initiatives hard to implement and maintain. When the ISO quality system initiative came about, it required them to define and document the day-to-day operations in their department and how they trained their employees and interacted with other departments. The need for these actions was difficult for some of the department managers to understand and was very different from how they were use to running their departments. However, these types of problems were not limited just to the lower and middle management level. We experienced difficulty in changing the mindset of a few of our upper level managers as well. While the implementation status was discussed almost weekly at our upper management meetings and there was verbal support for the system at an executive level, we continued to experience difficulty with some managers at various levels of the company fighting the system instead of letting it work for them. In fact it wasn't until a year after our certification that we had full "hands-on" involvement of the upper management group in our quality, "way of doing business" system. We are finally on the road to reaping the benefits that a dynamic quality and business system like ISO can provide.

—Anonymous

Managers Were the Tough Sell

Managers were the tough sell. We were never fully successful in conveying the concept to that level and above. This resulted in much last minute activity just before an audit thus diminishing the "quality" of the effort. The effort was more because it was required than because of a belief in the good and right of the system.

—Anonymous

When asked, What was the biggest disappointment? responses covered a wide variety of topics including:

- Management thinking that certification would solve all the company's problems.
- The unwillingness of managers to learn the standard. It was easier to let someone else do it.

It is disappointing how quickly persons became complacent and started allowing the system standard to slip.

—Yvette Castell, Quality Assurance Manager,
Dairy Industries (Jamaica) Limited

A big disappointment for those of us involved with ISO compliance has been the lack of understanding from customers and other food industry contacts of the full benefit and application that ISO 9000 compliance really means to our industry. It truly relates to the overall business and profit of an organization. Compliance does not guarantee quality, but it provides the structure and discipline for growth and improvement in every aspect of the operation. It is an important tool within the operation, but unfortunately, the level of understanding that we all hope for has yet to be achieved within our industry.

The biggest disappointment is how poorly ISO Registration is capitalized on by other departments within a company. The initial mindset is that ISO is a "Quality" initiative rather than a business initiative. This perception often prevents the company as a whole from leveraging the business decision to become registered.
—Jim Murphy, Manager of Design Process and Validation,
The Dannon Company

The biggest disappointment is in the fact that there are gaps in acceptance of ISO 9000 as part of our business culture. It is still seen by some employees as a separate entity that MAKES us do something that is not necessary for doing business.
—Sue Goode, ISO Coordinator, Cargill Corn Milling

Be Careful That Paperwork Doesn't Overwhelm the Efforts

The biggest disappointment experienced as a result of certification was that people get too caught up in filling out paperwork without asking what we are trying to accomplish. Many times the people felt that completing the forms had become the purpose.
—Peter Gottsacker, President, Wixon Fontarome

The entire certification process from implementation through certification assessment is a tough and stressful experience. There are those glorious moments of accomplishments and then those feelings of discouragement and disappointment. Many professionals commented that some of the biggest disappointments were a direct result of the certification process itself.

The biggest disappointment was the steady stream of postponements to the assessment schedule. Some were attributable to us; however, many were attributable to internal issues at our registrar. Trying to get employee enthusiasm and momentum for an initial assessment on the upswing three different times was a bit much.
—Dana Crowley, Production Manager, Danisco Sweeteners, Inc.

We were disappointed in the amount of time required to reach the certification phase.
—Henry Gibson, Quality Assurance Manager, Campbell Soup Company

Thoughts on Achieving Certification

The biggest surprises for us was how well we had done during the initial assessment. The process of going from no system at all to a matured system, was very painful for the plant. Due to this fact and ongoing reorganizations, we were a bit worried that we had missed some rather important aspects.
—Dana Crowley, Production Manager, Danisco Sweeteners, Inc.

The biggest surprise was actually being told that we were recommended for certification.
—Yvette Castell, Quality Assurance Manager,
Dairy Industries (Jamaica) Limited

Certification Has Resulted in Higher Standards

Compliance has meant that every employee is committed to rigidly following procedures and systems, resulting in a higher standards of process control and adherence to specifications. Compliance has also improved the quality standards of materials sourced from third party.

—Rex N. Gadsby, retired GM and CEO Dairy Industries (Jamaica) Ltd./Grace Food Processors, Ltd.

3
ISO 9001:2000—THE REVISION

First and foremost on our minds is ISO 9001:2000, the revision. What will it mean to us? What must be done to meet the requirements of the new revision? At the time of this printing, ISO 9001:2000 is scheduled to be published in the fourth quarter of 2000.

This revision has been developing for many years. Several drafts were released for comments prior to the final draft being approved. Certification cannot be officially granted until the new standard is published; however, registrars, consultants, and training organizations may assist companies in becoming compliant prior to that publication. Because of the considerable changes in format and fundamentals, registration to the ISO:9000 1994 series of standards will remain valid for three years after ISO 9001:2000's published date. This time period is meant to provide a sufficient opportunity for organizations to develop and maintain their system in a compliant and effective manner. However, it is recommended that organizations begin to bring their systems into compliance immediately. It will take some time to generate significant evidence demonstrating the system's full compliance.

The design and content of the new standard is meant to be easier to understand, have clearer terminology, and be more user friendly. It is also meant to provide a clearer connection between business processes and the quality management system.

The new standards will be

ISO 9000:2000, Quality Management Systems Fundamentals and Vocabulary

ISO 9001:2000, Quality Management Systems Requirements

ISO 9004:2000 Quality Management Systems Guidelines for Performance Improvement

ISO 9001:2000 will be the only compliance standard. Rather than having a choice of three (9001, 9002, 9003), from the 1994 Version' there will only be one ISO 9001:2000. ISO 9001 will define the requirements for certification of the Quality Management System whereas ISO 9004:2000 will provide a guidance document used to promote performance improvement. ISO 9004:2000 is not meant to be a guidance tool for interpretation of the ISO 9001 standard but to be used in conjunction with it. This guidance standard will focus on providing system guidance to extend the development of the compliant system beyond the requirements of ISO 9001:2000, focusing on achieving such excellence awards as TQM (total quality management) and the Malcolm Baldridge award for excellence. Mr. Tom Marchisello (Director of Quality Assurance, Campbell Soup Company) provides an interesting comment on this aspect:

> ISO 9001:2000—This revision of the standard appears to be taking quality management systems to the next level. It puts more of an emphasis on system performance and measurable improvement rather than simply compliance. It is taking the basic elements of Baldridge and applies the maintenance discipline throughout the certification and surveillance processes.

The newly designed ISO 9001:2000 standard is also more compatible with the content of ISO 14001 (Environmental Management Systems Specification). This is a positive improvement since now compliance to both ISO 9001 and ISO 14001 can be accomplished with little or no duplication or contradictory requirements. Terms are now addressed within the ISO 9001:2000 standard rather than in a separate sister document such as ISO 8402:1994.

Although compliance is now addressed in one standard rather than three, the ISO development committee did recognize that not all requirements of the standard would apply to all organizations. ISO 9001:2000 does provide "permissible exclusions"; however, the defined requirements for applying these are very specific. According to the Revised Standard, an organization will not be able exclude requirements that have an effect on it meeting customer requirements or any other related regulatory requirements. In other words, a company can no longer choose to omit its research and development department if in fact it plays a role in meeting its customer requirements. Permissible exclusions will be evaluated very closely by the registrars.

Jack West in the article "ISO 9000:2000 Shifts Focus of Quality Management System Standards" published in *Quality Progress* (October 1999) describes the transition in "quality," thinking that is reflected in the ISO 9001:2000 revision.

The original ISO 9000 family was based on the understandings of an earlier era in which quality was thought to be primarily a technical discipline—the purview of the quality professional. In the 1980s we discovered that this view was incomplete. We heard the slogan "Quality is a human resources problem, not a technical problem" so programs were developed to get workers involved in quality improvement. Today most organizations understand that all work is accomplished through processes, which are most effective when they are actively managed. As the 1990s advanced it became clear that quality has both a technical and a human side.

The revised version has an excellent focus. There is no doubt that as the principles and concepts are applied, it will bring quality management systems to the next level of effectiveness. Mark Atkins (President, ISO Quest Consulting Services, Inc.) has been working with some food companies with their development and implementation activities. Atkins provides the following insight on the application of ISO 9001:2000:

The impact of the year 2000 revision of ISO 9001 on food companies should be positive. The overwhelming criticism of the 1994 standard is that the list of 20 elements did not lend itself to "process thinking." Unfortunately, the documentation reflected this. Typically, under the 1994 revision, quality system procedures were written for each of the elements, often with much overlap and redundancy. The new standard's structure has been overhauled encouraging more of a process orientation and a reduced emphasis on element procedures.

Food companies should likewise develop documentation that is "holistic" and move away from the "fragmented" element approach. Currently, I'm working with a juice company implementing the new standard; this company had begun using the 1994 edition, but found it difficult to get buy-in from associates and management because they did not see the value in merely documenting the 20 elements. Our approach with the new standard is to first develop "process maps" for each of the core business processes. These maps serve as a business tool for training and as a means of "linking" to other processes and individual work instructions. Completing the process maps provides the foundation for meeting the requirements of the standard.

The interesting phenomenon with this approach is that employees (including management) had previously never understood how their role "fit" with other processes. They now have an understanding that enables them to identify non-value-added steps either eliminating them or identifying opportunities for improving their processes. Another benefit is that they are finding the [process] maps invaluable in preparing the company for a major expansion. They now have a tool for identifying constraints (bottlenecks) and the disciplined approach for adding equipment and personnel. Needless to say, there is no problem in getting buy-in now. The management representative is now a tremendous resource instead of an irritant. And guess what? They are finding that with some minor process adjustments, they are meeting all of the "shall" clauses of the new standard.

It is beyond the scope of this text to provide detailed step-by-step instructions on how to initiate development of or bring a current ISO 9001:1994 (9002, 9003) quality management system into compliance with the requirements of ISO 9001:2000. At this time, everyone is still going through a learning curve. The learning curve applies not only to systems but also to the registrars, consultants, and training organizations that specialize in providing and teaching the requirements. It is recommended that an organization carefully evaluate its sources of information. Be sure that training sources are credible. Confirm that registrars are providing effective training for its assessors and competent information on revision criteria. A transition document that provides guidelines for time scales and other implementation requirements should be provided.

As discussed in the Preface, knowledge for developing an effective and useful quality management system has been learned through experience with the 1987 and 1994 versions of the ISO 9000 series of standards. The content of this text is based on many years of experience. If you currently are approved to the 1994 version, it is important to have this system at its peak of compliance. This foundation will provide a meaningful and effective transition for the process improvement activities that will be required for compliance to the 2000 version. If an organization is not already certified, it is recommended that the information within this text be reviewed carefully and applied to the development process. Where appropriate throughout this text, comments and recommendations related to ISO 9001:2000 will be discussed.

The revised ISO 9001:2000 is based on eight basic quality management principles: customer focus, leadership, involvement of people, process approach, system approach to management, continual improvement, factual approach to decision making, and mutually beneficial supplier relationship. The standard is based on these principles; however, the principles themselves are not requirements. The basic requirements are defined in sections 4.0 (Quality Management System), 5.0 (Management Responsibility), 6.0 (Resource Management), 7.0 (Product Realization), and 8.0 (Measurement, Analysis and Improvement). Auditing will be against Sections 5, 6, 7, and 8. The "shall" statements in Section 4 are continued in the latter sections.

An effort has been made to highlight the requirements for ISO 9001:2000 in the related areas throughout this text. However, even this author, at this point in time, must rely on the experts (the developing committee) to communicate the revisions and explanations on compliance requirements. These "experts" understand what they are striving to accomplish. In doing so, based on their experience and expertise, they have provided the world of quality management brand new challenges. Initial efforts will rely on these sources for the training and guidance for application.

The text of this book is not meant to be all encompassing. It is based on many years of experience with the ISO standards; however, ISO 9001: 2000 needs time to acquire experience. There is no substitute for experience. We can study and learn, but we really won't understand until we work with it.

There is a lot to learn about ISO 9001:2000 with most of the practical learning only coming after we practice its application. We all hope that the learning curve will come quicker based on lessons learned from the previous ISO versions. The development committee has put many long years into its research and in reviewing the 30,000 plus comments and questions received on the draft copies. Their efforts have been commendable. Their main focus has been to provide the world of quality management the best possible management tool for providing customer satisfaction and continual improvement. It was great before, but now, thanks to their efforts and tenacity, it will be even greater. The 3-year "grace" period for compliance is fantastic, giving operations the opportunity to make the transition in an effective and useful manner. A transition document will be provided to aid all of us in making the transition. It will be up to us, the professionals and the users, to apply the knowledge and concepts to learn and grow with the maturing systems.

We, as professionals, should use the 3-year time frame efficiently and effectively applying the revised, but not so different, requirements to the development, control, and maintenance of stronger, more effective proactive quality management systems. It would not be appropriate to present information on ISO 9001:2000 based on experience when in fact the experience is yet to be acquired.

It is recommended that we apply what we have learned while preparing ourselves for the exciting challenges and learning opportunities with the application of ISO 9001:2000. Compare this with the experience and expertise required to be a brain surgeon. Having completed all the requirements to become a surgeon, he or she has the initial tools and knowledge needed. However, hard work and experience are needed to improve those skills. Given the opportunity to learn and grow within any profession improves everything we do. When I was a couple of years out of college and thought I had all the answers, a boss of mine told me that I just hadn't lived long enough to experience all the questions. No one has figured out how to teach experience. We just have to experience it.

There is no doubt that in applying the concepts and recommendations shared in this text your quality management system will have the foundation and structure to mature and grow through the application of the requirements of ISO 9001:2000. Comments and recommendations, as appropriate are included.

Thoughts on ISO 9001:2000 in the Food Industry

The food industry is gradually implementing ISO 9000:1994. The newer version of ISO 9000 has several appealing aspects, including the measurement of customer satisfaction. From a business perspective, I would want to know in concrete terms how I am meeting, exceeding or in need of improvement in the service of my customer base. The ISO 9000 standard provides you the opportunity to measure that. In quality, we see that if you measure something, it is going to improve. I see that the introduction of ISO 9001:2000 is going to advance the use of quality management systems registration in the food industry for that reason alone. Management is going to want some other objective measures that indicate their companies are doing the right things, and retaining customers.

—Brian Pugliese, Industry Relation Executive, NSF-International

Please note that all references to ISO 9001:2000 within this text is to the DIS/DRAFT version.

4
GETTING STARTED

In the beginning, it is important to get the program started and on the road to compliance. In doing so the program must be promoted as a "team project." The best pitcher on Earth would lose his motivation if every time someone hits the ball the fielders made an error. Although fans may always remember the ball that went through the first baseman's legs in the last inning to lose the game, in reality, there were innumerable plays throughout the game that resulted in the final outcome. One error doesn't make the whole game. When everyone on the team is doing the best they can, at what they do best, it brings out the best in everyone. Consider all associates within the system as part of the team. Let everyone take part. This empowerment brings everyone's knowledge and expertise to the program from the beginning. After all, the system belongs to everyone. What better way is there to get everyone involved then to empower the team spirit from the start?

THE EXPERIENCE. WHAT TO LOOK FORWARD TO

> In the fall of 1997, when my company began working towards ISO 9002 certification, my naiveté quickly became apparent . . . in short, I had no idea what I was getting into. I had management support, but that's all I had.
>
> I created a manual that I believed was so outstanding that the company's certification was guaranteed; I had heard other quality system people talk about how difficult certification was, but I, the novice, knew it would be nothing more than a brief inconvenience. I soon found out that, though beautifully written, fiction

has no place in ISO; our consultant taught me that. Immediately, the real job began.

It seems so obvious; quality system management in the food industry. Our company had been in business for years, and we knew what we were doing. All we had to do was "do what we say, say what we do, and prove it." How hard could that be. One hundred and fifteen employees soon made it very clear to me how difficult it would be.

Our company was recommended for certification in fourteen months, not something I would do again. Any company needs more time than that to build a strong system with sufficient documented evidence that proves that the company is actually doing what it says that it is doing. We struggled, fought, argued, cried, and became a company that pulled together for a common goal. In the early days, this was not the case. No one cared about ISO, or what it meant for us as a company or for our future in the industry. Everyone would look at me and say the same thing, "I don't have time." Everyone had too many responsibilities already. We were overworked, under staffed, and quickly loosing our sense of humor at the premise of taking on a new project. The word ISO was an expletive for a long time in our company. It took a lot of free lunches, donuts, awards for excellence in quality, and relentless conversation on my part for my colleagues, all one hundred and fifteen, to see ISO's value. (The ISO budget was very small so creativity on the part of the management representative is a must.) Very few times in my life have I ever been truly amazed, but this company's compassion and determination to make the goal of certification a reality will stay with me as long as I live. Even when I had second thoughts about the reality of a fourteen-month time frame, they empowered me with their strength of conviction, and belief in the system.

The worst part of ISO 9002 compliance: no one, no matter what they say, or how many times they have been through certification, can ever really prepare you for the rigors, hours, joys, frustrations, failures, and successes that come with it.

The best part: there is no greater satisfaction than seeing people you share a good portion of your life with (at least forty hours a week), commit themselves and their jobs to a process that many had never heard of in the beginning. They took ownership, their perseverance was rewarded, and their system lives on today, ever changing and growing as the scope of their business changes, and their commitment to quality remains. They belong to a very small group of food companies that committed themselves, their employees, and their resources to the improvement of the system they use to provide products to their customers. As is often said, "If it was easy, everyone would do it."

Today, I run my own retail company, and I often think of the team I was a part of, the change in culture that ISO brought to that company, and the great successes they have had as a result of ISO 9002 certification. The principles learned from ISO training and certification will always be a part of my professional life, regardless of the industry I am in. That's the true value of ISO.
—Ginna Young, formally Quality System Administrator, Louis Dreyfus Corporation; presently Vice President & Chief Financial Officer, Old Dominion Enterprises, Inc.

Defining your quality management system does not have to be as difficult as might first appear. Decisions and requirements will vary between operations and processes within the operation. First decide what is being done now, document the activity, keep records to confirm, then audit to determine compliance status and overall effectiveness of the activities being performed. Always focus on the fact that ISO requirements should be integrated into the present system rather than redesigning existing activities. Integrating is so important! If the requirements are not integrated, than the likelihood of duplicating efforts increases, which is not the best use of resources.

The actual amount of resources (i.e., time and cost) and the degree of difficulty necessary to bring a system into compliance depends on the existing process. Before beginning the activity, it is very important that those individuals who will be responsible for guiding the implementation process receive competent training in the ISO standard's requirements. Earlier it was stated that the standards are generic in nature and can be applied to any process or system. It is very important to have an accurate understanding of how the requirements relate to your industry. Research, benchmarking, acquiring the assistance of someone with credentials and experience in your industry will pay tremendous dividends in total cost, time, and employee motivation through all stages of the system definition, implementation, and maintenance.

What Were Some of the Critical Steps Necessary to Successfully Achieve Certification?

Russ Marchiando (Quality Systems Coordinator, Wixon Fontarome) stated: that they "took all of the proper steps by securing executive management approval, educating all of our managers, internal auditors and employees as to what was required to obtain ISO 9000 certification."

> Regarding the hardest part of the certification effort, if you don't have a true QC discipline in place, ISO will not be the magic "silver bullet" to get you there. You must have the fundamentals in place to implement ISO—otherwise the culture shock will be too great.
> —Peter Gottsacker, President, Wixon Fontarome

> One of the most difficult aspects of implementation was educating the staff on their role in achieving quality and how their job played a part in the even bigger process.
> —Ed Steven, Plant Manager, and Patti Smith, Quality Assurance Manager, NZMP

Making the Decision

Defining and implementing a compliant ISO 9001 quality management system is no easy undertaking. People say, "hindsight is 20–20." If ever it were true, it is true as one moves forward, learning with every forward step and every opportunity within the quality management system. Before we focus on getting started, it is interesting to hear what some would do if they could do it all over again with what they know now.

In Retrospect

If I started all over again with what I know now, I would require upper management unqualified support from the outset. Working around the periphery was a challenge. The spoken word, the written word, and the true feelings weren't in sync with one another. Saying quality, safety and production are equals is great. Sometimes they are ranked in that order, sometimes safety, quality, production was the order, but we never totally overcame the feeling among employees that production was really the silent number one. Employees hear what we say then see what we do and what we reward and spend time on. They read between the lines all too well.

—Anonymous

If I could do it all over again, I think I would have tried to make the organization of the system less complicated and limited some of the documentation at the start. But I did not know then what I do now! What we continue to do is to find ways to stream line the system and make it more efficient and useful.

—Sue Goode, ISO Coordinator, Cargill Corn Milling

If we could do it over again, we could ensure that managers and supervisors are totally committed, not just in words, but also in action. Marry the quality system into the day-to-day operations of the business, so that there is no differentiation between the ISO 9000 compliant system and regular business activities.

—Yvette Castell, Quality Assurance Manager, Dairy Industries (Jamaica) Limited

Grasping the idea of the quality system as a whole took time. Had we known better, we would have spent more time developing the base for the system rather than working on the parts and making those fit the system later.

—Sylvia Garcia, Environmental and Quality Control Manager, Domino Sugar

If I could do it all over again with what I know now, I would start the process with a simpler scope.

—Gail Cartwright, Assistant to the AVP Human Resources Department, Bacardi & Company Limited

I was lucky enough to have this (second) chance so I can answer from experience. The key learning from my first ISO experience which proved invaluable the second time around was simply start the system immediately. Do not wait to "go live" until the system is perfect. It makes no sense to disable a continuous improvement system until it is error free. This is similar to choosing not to use a self-sharpening saw because the blade is not sharp. If you use the tool the tool will improve with use. I suggest establishing a few essential procedures and letting the system do the rest. Examples of some of the activities that promote this are, document and data control, which is needed to provide consistency and yields a master list of required documents; internal audits [that] provide a

mechanism for detecting gaps in the system; and corrective/preventive action that provided the mechanism for continuous improvement.
—Jim Murphy, Manager of Design Process and Validation, The Dannon Company

I would make the system much more flexible and really ask what does the customer want? What is the customer willing to pay for? What is our management willing to commit to the design? I find our organization spends a fair amount of time interpreting the standard rather than answering what is it that our business needs. We tend to find ourselves creating "systems" that our management does not want, our workers do not want, and our customers don't know or care about. Then months down the road we wonder why continuous improvement and customer satisfaction are not moving after all the investment.
—Anonymous

We are a fairly large facility with many varied departments. In striving to install systems to meet and exceed the ISO requirements, many efforts were departmentally based. We began to have many small ISO 9001 quality processes. We later unified into a company wide quality system with many departmental responsibilities. If we could start over, we would put forth a greater effort into establishing these systems company-wide from the beginning.
—Keith Gasser, Quality Systems Manager, Tropicana

If we had the opportunity to "re-develop" the system, the first and foremost task would be to build in economic measures for each functional requirement of the standard as it was developed and implemented. This would facilitate economic justification and improvement while including all the elements of a recognized quality system. We would also reconsider the level of detail to which we took the system as there are some opportunities to streamline it and make it more functional to the business. The good news is that the system is in and we can go back and improve it in those terms.
—Mike Burness, Director of Quality Assurance, Pepperidge Farm, Inc.

Timeline and Culture Change

At the onset of the implementation process, it is important to be realistic when identifying "timelines" and resources. Do not be over aggressive.

If I could start the whole process over with what I know now, it would start with the initial timeline of 18 months we targeted for certification. After going through the implementation and certification processes and achieving our goal of 18 months, it was not easy. The culture of the company needed to change as the ISO implementation and certification processes proceeded and unfortunately the culture change was much slower than we anticipated. This was especially true at the middle and lower management levels. As the timeline moved closer towards certification, the QA Department realized that if the certification were to succeed that they must do what it takes to accomplish the goal. Fortunately, we did reach the goal but unfortunately we not only had some QA personnel near "burn-out" but we also had a quality system that was built primarily by QA with some input from department managers. This ultimately posed a problem for the growth of the quality system after certification because it gave the department managers an excuse for being detached from this QA department-imposed quality system. It was not "their" system and so their willingness to be involved in the growth and maturing of the system and the resultant accountability to the system were not present. All of this happened despite training all managers from all levels and internal auditors as to the requirements of ISO 9001 and educating all employees on the basic principals of ISO 9001 and what it would mean to them. We also had executive management support at the onset. However, it became clear that what they needed to do to support the implementation and certification process was not quite understood by either them or QA.

—Anonymous

Maintain the Focus

Maintaining the focus. Being a fairly large facility, it was a three-year effort to become certified. Companies and personnel change over three years and we had to change and adapt to these changes while maintaining the focus. To maintain the enthusiasm for such a long time period, we had to establish several milestones along the way. It was very important to celebrate accomplishing each of these milestones.

—Keith Gasser, Quality Systems Manger, Tropicana

Thoughts on the "Extraordinary" Effort Required

One of my biggest surprises with the implementation process was the extraordinary amount of time, effort and attention to detail required to achieve certification and the amount of time necessary on an ongoing basis to keep the system current and maintained.

—Dave Demone, Environmental and Quality Control Manger, Domino Sugar

The biggest surprise was the amount of work that was required to get certified. The biggest disappointment is finding how not ready we were after the first preliminary assessment.

—Gail Cartwright, Assistant to the AVP Human Resources Department, Bacardi & Company Limited

The biggest surprise, as any other large company might agree, was that many system requirements were already documented and in place from basic business operation. Items like management review (in various forms), purchasing, process control, traceability, etc. are basic business functions in most companies. The surprise came when we learned not only that we had these systems in place, but that implementation of the ISO requirements could actually improve them. Documenting more specific requirements for purchased product, for example, is an area that we improved due to the requirements of the system.

—Mike Burness, Director of Quality Assurance, Pepperidge Farm Inc.

One of the greatest surprises was the identification of a number of processes that had been in existence for some time and as it turned out were actually a

hindrance to our operation. These were costing us money and dragging our team's efforts down.
>   —Jon Porter, President, J. Porter and Associates, Ltd.

I was surprised at the amount of time it takes to devote to the process of becoming certified.
>   —Charlie Stecher, Quality Assurance Manager, Reckitt & Coleman, Inc.

Choosing the Team

When getting started, it is recommended that an ISO implementation team be created. Some organizations choose a team comprised of top management who act as the steering committee providing support and resources. A second team made up of representatives from the various system areas led by the management representative may be identified to perform the hands-on implementation. Others create implementation teams for each department. Whatever means is chosen, keep in mind it is essential to the system's ultimate success that implementation be promoted as a team program with total support from executive management.

The Team

The importance of each manager understanding his or her responsibility and taking an active ownership for his or her related activities cannot be overemphasized. The management representative or the ISO implementation team should not "overindulge" by shouldering the responsibilities for all areas.

Russ Marchiando (Quality Systems Coordinator, Wixon Fontarome) states that if he could do it all over again with what he knows now, he would take more of "the role of a consultant providing information as needed letting middle management develop a system that would be 'theirs.' By compelling them to develop their portion of the quality system they would also take ownership of the system realizing that no one else is more responsible for their areas than themselves."

> Many of the ISO certified companies that I have been involved with state that if they could do it all over again with what they know now that without a doubt they would spend more time in selecting the people who are going to work together to complete the system. People skills were not considered in the primary selection and it became an issue. They also stated that they would be sure to allow sufficient time for the implementation phase.
>   —Jon Porter, President, J. Porter and Associates, Ltd.

> Decisions have to work for your people. It is important to believe in your team and to recognize them whenever possible. With our certification we were all very proud that our team was recognized for their accomplishments in the global market place.
>   —Linda Taylor, Training and Quality Manager, Le Meridien Jamaica Pegasus

Understanding the Standard

> The ISO standard is hard to read. The wording is meant to be generic in nature, but this sometimes makes it hard to understand what it is really saying such as the "contract" referring to a customer order.
> —Bill Lockwood, Package Quality Manager, Hiram Walker & Sons, Ltd.

Internally, there must be a strong understanding of the ISO requirements. Depending on the size of the implementation team, one or more individuals should attend an accredited ISO 9000 Lead Auditor course presented by either a registrar or an accredited training or consulting firm. It may not be economically feasible to have the entire team attend the Lead Auditor course; however, other team members will benefit from an internal auditor training course. This is usually a two-day course focusing on understanding the ISO standard, auditing requirements, and auditor protocol. This will give the team members an excellent background for system application.

Communication

An ISO familiarization meeting should be attended by every associate, designed specifically for your operation. This meeting is usually scheduled for one to two hours and includes a basic overview of what ISO is, the implementation process, what certification will mean to the company, the importance of team involvement, and every associate's role in achieving this. Also, as a precursor to beginning implementation activities, a letter should be written to all associates from the executive manager or management team providing a brief explanation of what is to come. This provides the "kick-off" for implementation and total associate involvement through team activities from the (very) start!

Thoughts on Communication

> The hardest part of the implementation process was getting the employees to be consistent with the documentation. For the most, this was a result of poor preparation . . . and lack of total communication. Employees did not fully understand the implementation process and the role that they played in achieving certification. Sufficient time must be spent in presenting the material, purpose and "results to be gained" to everyone in the establishment.
> —Jon Porter, President, J. Porter and Associates, Ltd.

> If I were to do it again, I think I would get more people involved early on to understand what the whole concept is about and to better understand the overall process. I had sat in on a few meetings during the initial phases and initially was too confused to understand what was involved with developing a quality system and what it meant for a manufacturing plant.
> —Rick Bay, Plant Manager, Reckitt & Colman, Inc.

The Product

The system's "product" must be identified. For example, for a manufacturing location, the product is that item that is being manufactured. However, this decision is not quite as easy to define, if in reality the product is a service. If the company's business is to perform calibration services, then the product may be defined as "the service of performing calibration." That indeed is what the company does, thus that is the "product." The process for performing that service will then be the system that is defined, implemented, and certified to the ISO requirements.

If a corporate office decides to certify their purchasing department, then the product will be the service of purchasing. Requirements for performing that service would be addressed through "process control." This can be confusing since there are requirements in the standard regarding the "purchasing" function. For a purchasing department this would, however, reflect its suppliers of "quality-critical" supplies and services related to producing its product: the "purchasing service."

Thoughts on Defining the Product

> One of the challenges we faced early in the process of creating our system for obtaining ISO 9001 certification was clearly defining what our product was. In our function as both a research and a testing laboratory, we concluded that our product was information, which came in the form of data, advice, formulas, and processing procedures. Once we clearly understood what our product was, we had to define what measures were necessary to ensure the quality and integrity of the information the lab generated. After that, it became easy to implement and document our procedures.
> —Victor V. Margiotta, Director of Quality, SOBE

Tom Marchisello (Director of Quality Assurance, Campbell Soup Company) stated that being a corporate function presented a challenge in

> interpreting the standard for a service organization within a manufacturing corporation, and making our specific work compliant. It was also difficult identifying our key internal customers and defining their requirements.

Where to Start

The answer to the question, "how do we know what to do first" is critical to the process. Compare it to planning a road trip. If an individual who had never left New York City planned a trip to Phoenix (Arizona), it would be essential before leaving home to have a road map. An accurate road map can make the difference between a fun and successful trip or a very long and wearing experience. With this thought, let's think about road mapping the ISO implementation process.

More Decisions

The decision to seek certification is the first decision. Subsequent decisions such as choosing a training organization, a registrar, and whether or not to use a consultant will also need to be made. Base these type decisions on a conscientious review of credentials. Benchmarking with companies with similar type products is an invaluable source of information. Always ask for credentials and recommendations. Registrars and consultants should provide references and company approval listings of those with which they have worked within your industry.

Deciding on a Consultant

There are many consultants that would be happy to assist with this project. The decision as to whether or not to seek assistance from a consultant must be made wisely. Consultants should provide evidence of their expertise in your industry. The consultant must be effectively experienced and able to translate the generic basics of the ISO standard into requirements for your specific industry. Ask others in your industry for suggestions. A consultant that has experience in working with applying the ISO standard within your specific industry can be invaluable.

Yvette Castell [Quality Assurance Manager, Dairy Industries (Jamaica) Limited] once shared a quote from her general manager (Mr. Cameron Bisland): "If the company team has the assistance and guidance of a competent and experienced consultant than it will have a system that will withstand the challenge of any audit."

Checking references and confirming credentials is worth the time. Remember there is no substitution for experience. Knowledge of ISO is only the beginning. This must be combined with knowledge and experience in the related industry. Interview and ask questions. Meet with a perspective consultant personally to determine if he or she will fit within your organization. As an essential team member, the consultant must have the personality and expertise to work with your team in an effective manner. The key is to work with your team.

Be careful with consultants that want to move in and do it all. The true and effective role of a consultant is as a guide. If the consultant writes the program, then it will be her program not yours. The consultant should provide guidance to ensure that the system is defined and implemented in the most effective, economical, and useful manner as possible.

Thoughts on Choosing a Consultant

> We were not experts by any stretch of the imagination and trying to decipher a document like the ISO standard gave us fits and led to overcompensation. Having a more experienced eye would have helped out a lot. Initially, we read

so many things into the standard that really weren't there. As a result, our earlier procedures were so cumbersome that they were unmanageable. Even today, there are a few aspects that I know we went overboard on. This single-handedly dragged out the initial certification process an additional year. In hindsight, I feel that we should have been better served if we used our consultant more to our advantage. Our consultant was very good but also expensive. Thus the reason we did not use too much consulting. In retrospect though, utilizing consulting would have saved time and money in the long run. I know this can be a touchy subject because in many cases, consultants tend to be self-serving and make you go overboard to make sure that you do get certified. However, the system that they leave you with may be more their system than yours, becoming very difficult to be maintained 2 years down the road. Clearly, one has to pick a consultant that has a good track record and comes highly recommended from reliable sources.
—Dana Crowley, Production Manager, Danisco Sweeteners, Inc.

When interviewing your consultant, discuss his or her philosophy for "assisting" your efforts. A consultant's role must be to provide guidance and enhance the understanding of the ISO standard's requirements not to define and "create" your system. It is your system. The requirements must be integrated into the system, resulting in a quality management system that is not only useful but provides a sound foundation for continuous improvement. Think of the analogy of practicing to become a concert pianist. One can get Beethoven, Chopin, and all the great pianists to teach the fundamentals and provide guidance through the learning and practicing phase. But actual achievement is only reached when it is your fingers and mind actually playing the piano with the expertise of a concert pianist. No one can do if for you. If someone did, then it would be his or her concert not yours.

Rick Aldi (Director, Quality & Environmental Affairs, Hiram Walker & Sons, Ltd.) stated that he would rely less on his consultant to "lead ... to victory and more on the training provided by the registration body." When Ed Steven and Patti Smith (Plant Manager and Quality Assurance Manager, NZMP) were asked "what would they do differently if they could do it all over again?" They responded that they "would not have used inexperienced consultants in the beginning."

If we had to do it over again with the knowledge we have, I would hire a consultant to assist us in defining the process and facilitating our efforts. We wasted 6 months going the wrong way and doing the wrong things.
—Charlie Stecher, Quality Assurance Manager, Reckitt & Colman

If I had it to do over again I wouldn't change the implementation process, as the process used, involved every employee with total commitment coming from the top, the Board of Directors and Senior Executive.
—Rex. N. Gadsby, retired GM and CEO Dairy Industries (Jamaica) Ltd./Grace Food Processors, Ltd.

Returning to the road map comparison, the consultant will not drive the car but will provide the guidance and expertise needed to make sure that you don't drive from New York City to Toronto on a path to Phoenix. It is the consultant's responsibility that the road map lead south toward Phoenix and not north to Canada. Without this guidance (road map) one might very well end up in Toronto before realizing something is wrong. There are many books and guidance materials available, but remember, depending on the industry, there could be those four-way intersections that require "experience" to make the most effective decision. It is in this role that the consultant experienced in understanding and relating the requirements to your industry will be an invaluable and cost-effective resource. Keep in mind that it isn't the end of the world if the system takes a detour, however, too much of this will drain resources and be demotivating to the team's enthusiasm.

Choosing the Registrar

A registrar is hired to provide the certification service. An auditor or assessor working under the direction of the registrar actually performs the certification functions. The choice of which registrar to use is as important as any other decision made during this process. Perform constructive research to identify those with experience in the specific industry. Be sure and ask for proof that the registrar is accredited in your industry. Regarding the food industry, there are many that would like to, but few that have long-term experience and familiarity. When evaluating the registrars, ask for and check references, request backgrounds and resumes for the assessors, ask for a sales visit, and whenever possible, meet the assessor before making the final decision. The registrar should promote his or her role as a team member to your system. Although the registrar cannot consult and instruct on how to define your system, he or she can provide a wealth of support information based on experiences and expertise that will be invaluable to your process.

Thoughts on Choosing the Registrar

> Ask prospective registrars for the name of other companies that they certified that are in your industry. Ask for contact names and numbers, not only to know how well the registrar served them but to also benchmark against a successful company. If there are no other companies like yours that are certified, or perhaps just a few, think of yourself then as a pioneer and being one of the leaders in your industry to take this "quality" step forward. Ask to meet or at least speak to the assessor on the phone that would most likely perform the assessment. Request information on the assessor's background and certified companies that he or she has worked with. Ask them specific questions about the industry to ensure that their background will meet your needs.
> —Charlotte Sladek, Business Development Representative, LRQA

Food companies registering their quality systems should seek a company that understands their industry and that understands food safety. We too often hear that food companies seeking ISO 9000 registration select registrars that have no food industry specialty. Confirm that the auditors have the proper "codes" (SIC or NACE) required to conduct the audit. Even having the proper codes, the auditors may not have degrees in the sciences, day-to-day experience with the design and implementation of Good Manufacturing Practices in a food setting or understand the principles and applications of Hazard Analysis Critical Control Points (HACCP). Disappointment occurs during those phases of the audit when the quality assurance staff needs to hold miniworkshops to explain fundamentals of GMPs or prevention-based food safety systems, like HACCP to the auditor. In short, a food company needs to verify the core competencies of the auditing company and of the auditors themselves to assure themselves of a smooth, hassle-free audit.

—Brian Pugliese, Industry Relation Executive, NSF-International

More Thoughts on Choosing a Registrar

There are many factors that companies consider when choosing an ISO 9000 registrar. Key attributes typically evaluated include: experience, accreditation, reputation, administrative support, and price. More intangible attributes also gaining popularity in the selection process are related to the supply chain, auditor training, and the combination of auditor qualification and geographic location. More and more companies are reviewing the supply chain to identify those who have registered key suppliers and customers. It is felt that evaluating results from key suppliers can be a source of objective feedback on the registrar's performance. Also, a review of key customer registrations may reveal a common registrar with whom it may be beneficial to work with in obtaining certification. The feeling may be that "if our customers use them, then there will not be an issue of recognition when it comes time to submit the next bid."

Another area of increased consideration is auditor training. Training, not necessarily in the area of audit experience, but the type and extent of training the registrar provides to the auditors. What type of initial and continuous training, if any, is provided? How is the registrar's audit philosophy relayed to the auditors? How is it reinforced over time? How does the registrar maintain auditor continuity and consistency? These are just some of the questions that are being raised related to auditor training.

Finally, another factor that some companies are placing emphasis on is the combination of auditor qualification and location. Highly qualified auditors who are also locally based are increasingly in demand by companies. Minimizing travel expenses is an important consideration. Some companies may resort to changing their registrar if they can find a local auditor with similar qualifications and attributes available with another registrar.

Whether tangible or intangible, these attributes are becoming increasingly more important to companies evaluating and selecting a registrar in order to find the most suitable business partner for their ISO 9000 endeavor. There are plenty of

variables to consider and the emphasis on one factor versus another will depend on each individual organization and its priorities.
—Darren Weber, Business Development Manager, SGS International

Weber's comments are extremely applicable in this age of soaring costs. Unfortunately, local availability for qualified food assessors is very rare. It is important to check out the complete criteria, but keep in mind Pugliese's comment that a food company does not want to be put in a position of having to train the auditor in basic food industry requirements. Do not sacrifice experience for cost. Be open and frank with the registrars. Discuss your concerns. A reputable registrar will be open and honest in the negotiations with you. After all, remember the registrar will be part of your team. Honesty and commitment must be a primary attribute of any team member

The Gap Analysis

A gap analysis is an important activity performed early in the implementation stage that focuses on comparing existing activities to compliance requirements of the ISO standard. The gap analysis identifies the "gaps" between what exists now and what must be present to achieve certification. This can be a very enlightening experience and if performed effectively, very reassuring. It is common to discover that existing processes and activities are closer to compliance than originally thought.

Who Should Perform the Gap Analysis?

Results from this analysis are important in providing the road map of what must be done. The gap analysis could be performed internally. However, performed by a person external to the system, such as an auditor or consultant, would provide an independent system baseline based on his or her experience applying the ISO standard's requirements to similar processes. Some registrars offer this service. It is not considered consulting because the report is a statement of fact not a statement of recommendations. A consultant may be in a position (based on his or her experience and expertise) not only to perform the gap analysis but to provide guidance to bring the process into compliance status.

Preliminary Assessment

The preliminary assessment (also referred to as the "prelim") is just what the name implies, a preliminary ISO 9001 system-compliant assessment performed prior to the actual official "main" assessment. Many refer to this as the dress rehearsal. Depending on how it is contracted with the registrar, it may be performed with a focus on the whole system or on various areas of concern. Every effort should be made to have the registrar assign the same assessor, and, depending on the length of time for the assessment, the same audit team that will be performing the main assessment. This gives management and associates the opportunity to meet the assessors and hopefully gain a degree of comfort. It also gives the audit team a chance to get to know the process and the associate team. If performed effectively, this will build a sense of comfort and familiarity between both the auditors and the auditees.

Thoughts on the Preliminary Assessment

Tom Marchisello (Director of Quality Assurance, Campbell Soup Company) stated that one of their

> biggest surprises during the implementation process was that although overall they felt that they had an effective and compliant quality management system in-place, there were still many noncompliant areas identified during their preliminary assessment. Stress the fact to management and the implementation team that the preliminary assessment, [the dress rehearsal] is an essential part of every ISO implementation process.
>
> The biggest surprise that we experienced was the amount of work that was involved in the process. It took us two preliminary assessments, but we made it and it was absolutely worth the effort.
> —Andy Fowler, Research Engineer/Management Representative, Bacardi & Company Limited

Certification Process

In actuality, the certification process begins with the expression of interest and contacting the registrar of choice. Services may vary between registrars. However, most include the option of a preliminary assessment, followed by document review, initial or main assessment, surveillance visits, and triennial reassessment. In choosing a registrar, it is important to meet with those being considered and to understand their program and requirements. Weigh the considerations as they best relate to your specific system. Check references, and companies that they have approved. Always focus on what is best for your specific system. The team must be confident that the registrar will provide as efficient and economical service as possible.

Document Review

Most registrars perform this as an on-site visit prior to the main assessment. The purpose is to review documentation (i.e., the quality manual) to confirm that all ISO 9001 requirements have been addressed. Ideally, the lead assessor performing the main assessment will conduct this visit. If an operation has not had a preliminary assessment, then this would be the "get-acquainted visit" to confirm assessment criteria and scope. The program for the main assessment is usually created during this visit. Some registrars perform the "document review" as a desk audit off-site, but again team members tend to be less nervous during the main assessment when they have had a chance to meet their assessor prior to that visit.

The Main Assessment (Certification Audit)

The registrar will assign a certified lead auditor and where appropriate an audit team to perform the main assessment. The purpose of this assessment is to evaluate the system and confirm its ISO 9001 compliance. If no major noncompliances are found, certification will result. Major noncompliances are defined as system failures. These require an additional visit to evaluate and confirm that the corrective action taken brings the system into compliance. Depending on the time period that may elapse between the main assessment and the evaluation of the major noncompliances, a complete system reassessment may be required.

Should no major issues (system breakdowns) be identified, then the audit team will recommend the system for approval based on its documented confirmation of system compliance. The approval certificate will be issued directly from the registrar's office once its management reviews and confirms that the audit report provides significant evidence to demonstrate compliance. Registrars themselves are audited carefully by the accreditation agencies to ensure that its activities are in compliance. In order to maintain its accreditation status, the registrar must perform according to specific requirements and maintain records to demonstrate compliance to these requirements.

Surveillance Visit

A certificate renewal visit, also known as a triennial reassessment, is usually required every 3 years. During this 3-year period the registrar will assign an assessor to perform partial on-site evaluations of the system at a frequency defined in your contract. As a general rule, this time period is usually every 6 months throughout the certificate's life. Some accredited registrars offer annual reviews. Although some systems do quite well on an annual basis, immature and/or newly implemented systems seem to mature much more effectively with the discipline associated with the semiannual review. This must be an internal decision. The decision should be based on what is best for the specific system and not what might be easier or more convenient. It might be tempting to avoid the inconvenience associated with frequent external visits, but remember this system is being developed as a useful, meaningful tool to promote continuous improvement. It is not maintained because the external certification registrar evaluates it every 6 months. It is maintained for the structure and discipline that it brings to the entire process.

The surveillance visit will include an evaluation of the management review, internal audits, corrective/preventive action processes, quality system documentation revisions, and outstanding noncompliances from previous visits. As time allows, over the 3-year life of the certificate, other areas may be evaluated; however, emphasis is placed on confirming what is being performed internally to maintain the system. Once the system is confirmed compliant, it is up to system management to "maintain" the system compliant to the requirements of the ISO standard.

The compliance certificate is generally issued for a 3-year period; however, registrars may vary with their accredited methods for certificate renewal. In most instances, certificate renewal will require a complete system assessment such as that performed during the main assessment, but again this may vary between accredited registrars.

Certification Versus Just Applying the ISO Philosophy?

Can the benefits from an ISO-compliant quality management system be experienced without actually going the extra step and expense to obtain third-party certification? There are many opinions on this, with the majority feeling that the extra discipline of an outside assessment makes the system stronger.

Bill Lockwood (Package Quality Manager, Hiram Walker & Sons, Ltd.) stated that "we could try to keep the system in place and eliminate the ISO certification, but a reality check is needed, as we are all human. Without the driving force to stay ISO certified, we may not be as disciplined to stay on track."

Many have mixed feelings on the benefits of actual certification.

> At times I wonder if it was worth the effort to obtain certification. There is little doubt that adopting the ISO 9001 philosophy was worth the effort but I am not convinced that taking the next step towards certification was worth it. ISO 9001 certification is still not very prevalent in the food industry and I am not convinced that there is strong understanding of what certification entails and how it applies to the quality of product and service that we can provide for them. Certification is an expensive endeavor and I am not quite sure that it has paid off as opposed to adopting the ISO 9001 philosophy and not obtaining certification. However obtaining and continuing certification provides the corporate accountability to the quality system and helps to maintain the urgency to keep the system current and compliant.
>
> —Russ Marchiando, Quality Systems Coordinator, Wixon Fontarome

> First let me state that ISO is worth the effort in terms of improved business systems (efficiency, quality, and customer focus). I do not believe that the value is necessarily a function of certification. I strongly feel that an ISO "compliant" system offers as much value as an ISO certified system in most companies. Unless the company has a marketing opportunity which can leverage ISO registration I can see no incremental value of hanging a certificate on the wall.
>
> —Jim Murphy, Manager of Design Process and Validation, The Dannon Company

In the end I think that getting certified was worth the effort, although, there are certainly times when I have questioned this. There's no doubt in my mind that the ISO 9001 standard provides a great framework for driving continuous improvement throughout the organization. While you could apply these same principles to your organization without going through the certification process, the likelihood of successfully implementing and maintaining such a philosophy in our organization would be extremely difficult without 100% management support. I'm not saying it couldn't be done but the certification process and the ongoing upkeep of the system provide deadlines and urgency to keep the system up-to-date that normally wouldn't be there. While the certification process is expensive, the quality system effectiveness as judged by external auditors, along with the feedback they provide, keeps us moving in the right direction. The certification and periodic surveillance audits, in addition to the significant financial resources committed to this process, provide the leverage needed to maintain and grow our quality system.
 —Tim Sonntag, VP Quality Assurance & Technical, Wixon Fontarome

Stick with It

Many professionals have shared with me that once the system is implemented it becomes evident just how extremely worthwhile the efforts had been. In my experience, sometimes management uses premature judgment on the value of the process and frankly "give up" before completing its implementation. One company that I was working with had stopped working on certification at the direction of their corporate group. After several "avoidable" problems, management made the decision to reinstate the project. Their stated focus was to apply the structure and discipline inherent with ISO 9000 compliance in a manner that would provide a foundation for continuous improvement. They gained corporate approval by using the research and collection of historical process data to justify their efforts. Once the implementation was complete, even prior to the actual certification, evidence of overall process improvements enhancing both quality and food safety activities had become apparent. Corporate management praised the operation and requested all other company owned facilities to begin implementing the requirements.
 —Jon Porter, President, J. Porter and Associates, Ltd.

Just a Few More Thoughts from the Author

Before the subject matter moves forward on specific compliance activities, it may be appropriate to share a few more thoughts:

Remember the concept of the relay race. An auditor will ask questions such as "How do you know the product you receive is acceptable?" "Where does your responsibility begin?" Where does it end?" Associates do not need to know other's responsibilities, but they do need to understand where theirs begins and ends.

Keep in mind that within the ISO system, basically everything really relates to something else. For example, purchasing requirements may relate to shipping activities for the use of approved suppliers of shipping services. Inspection and testing activities may relate to requirements of calibration. Document control, quality records, and training relate to every other element of the standard. Ensure that overlapping responsibilities and requirements are appropriately addressed in each area. Specific procedures and/or work instructions should be readily available in areas that need to access information and requirements contained within these documents. Ensure that associates are trained on the documents with records maintained confirming the training.

ISO compliance should become a way of life for everyone throughout the system. Be very careful that compliance activities do not become the responsibility of one person or one group that polices everyone else.

When defining the requirements be careful not to include "uncertainties" or leave anything to interpretation. Do not use words or phrases that leave interpretation of their meaning to the reader. Examples of these would be "as appropriate," "as necessary," "periodically." The meaning or translation of responsibility for these types of terms basically depends on the reader.

Be careful with assigning responsibilities for requirements required by the ISO standard to areas outside of the scope of approval without clearly defining the control executed within the scope. An example of this would be if the corporate office performs purchasing. It would not be appropriate to make the statement that "the identification and evaluation of approved suppliers is the responsibility of the corporate purchasing department" unless, of course, an audit visit to corporate purchasing was planned during the certification audit. The following statement would define the organization's responsibility in meeting the requirements of the ISO standard.

> Although the corporate purchasing department identifies and evaluates the suppliers, our system actually verifies that the supplies meet the requirements through incoming inspections. Nonconformances and supplier performance issues are reported to corporate purchasing. Our system through inspection and testing has the final decision as to whether a supplied item is fit for use in our process.

Remember the records to demonstrate that these activities are being performed must also be identified and maintained as quality records.

Be careful when defining frequencies. Does once per year or once annually mean once per calendar year or every 12 months? Also watch the use of terms such as biweekly or biannually. Some dictionaries actually give them dual meanings. "Biannually," according to some dictionaries means twice a year or every two years. Two very different meanings!

In Conclusion

Defining the related ISO standard requirements does not have to be as difficult as might first appear. Decisions and requirements will vary between operations and within processes within an operation. First decide what is being done now, document the activity, keep records to confirm that the activities are being performed, then audit to determine compliance status and overall effectiveness.

Never lose sight of the fact that ISO requirements must be integrated into the current process rather than redesigning existing activities. If you do not integrate, then you will duplicate. Duplication will deteriorate resources and dilute the system's effectiveness.

5
QUALITY MANAGEMENT SYSTEM (ISO 9001:2000 SECTION 4.0)

5.1 AN OVERVIEW

The requirements of ISO 9001:2000 Section 4.0 basically corresponds with ISO 9001:1994 Section 4.2. The draft standard states: "The organization shall establish, document, implement, maintain and continually improve a quality management system in accordance with the requirements of this International Standard." (ISO 9001:2000) (draft) Section 4.0). This section begins to introduce the process-based approach that addresses the connections (interfaces) between system operations. The system must be defined in a manner that more clearly reflects the needs of the organization as it focuses on daily operations. It must focus on continual improvement, preventing nonconformances, and meeting customer needs and requirements. To assist in an organization's transition, those systems that had approved to ISO 9001:1994 may choose to rewrite their quality manuals in accordance with the revised style of ISO 9001:2000; however, this is not a requirement.

Less emphasis is placed on specific documentation. Regarding the required documentation, the draft standard states: "The quality management system documentation shall include: documented procedures required in this International Standard [and] documents required by the organization to ensure the effective operation and control of its processes" (ISO 9001:2000 (draft) Section 4.0). This provides the organization the opportunity to choose which documents it will require. However, should it be decided not to use documented procedures other than those required by the standard, it will have to be demonstrated that alternate methods are being maintained to control the process.

This text recommends documented requirements in procedures and work instructions. As will be evident through input received from the field, defining requirements is important to the overall effectiveness of the quality management system. Having requirements defined and documented provides structure and discipline while decreasing the opportunity for variation. In preparing the documentation, be cognizant of the interfaces between the processes. It will be very evident that when applying ISO requirements "everything relates to everything else." The revised standard now provides the organization the opportunity to map its process, defining the interrelationships in a manner that is most useful to the specific system.

5.2 QUALITY SYSTEM

The term "quality system" is defined in ISO 8402:1994 as "organizational structure, procedures, processes, and resources needed to implement quality management." The food industry quality system should address relevant codes of practice and legal requirements such as weight control, GMPs (Good Manufacturing Practices), HACCP (Hazard Analysis Critical Control Point), and standards of identity.

Thoughts on the Quality System

> Certification provided us with a system which was all encompassing and the best thing was that most employees saw it that way. The quality policy or mission statement provided the purpose. Policies and management procedures provided the skeleton in which it rested. Operating procedures provided the tools to allow us to meet the purpose while the quality audits provided the means to evaluate the effort. This all resulted in continuous improvement.
> —Sylvia Garcia, Environmental and Quality Control Manager, Domino Sugar

> With our quality management system in place, we have been able to tackle new challenges head on, and use the process to effectively implement the required standards. It has stood firm through rebranding, restructuring, change in General Manager and a most challenging business environment.
> —Linda Taylor, Training and Quality Manager, Le Meridien Jamaica Pegasus

The *Food Guidelines* (1995) states that

the quality system should ensure that all those activities within the organization that could impact on the quality of the product are consistently documented. Above all, the structure of the quality system should be right for the company. (p. 5)

The ISO 9001 standard has specific defined requirements related to defining, documenting, controlling, and maintaining a quality management system. These requirements must be clearly addressed in the system documentation. This begins with the quality manual that will define the structure of the documentation and cover the requirements of the current ISO 9000 standard.

The Quality Manual

The quality manual is designed to address all the requirements of the ISO standard. It can be generic by restating the "shall" statements as "will" statements. For example, the ISO standard states "management with executive responsibility 'shall' review the quality system at defined intervals." (Note that ISO 9001:2000 now uses the term "top management.") In response, the quality manual may state that "management will review the quality system at a minimum of quarterly" or "management will review the quality system at intervals as defined in Procedure AB-01." Notice the replacement of the "shall" with "will."

The certifying registrar will evaluate the quality manual to confirm it addresses all the requirements of the standard (all the "shall" statements). This evaluation will be done prior to actually performing the compliance or main assessment. The quality manual must link or make reference to related procedural level documents where the requirements are defined. It must also outline the structure of the quality system documentation.

The Documentation

The establishment and maintenance of documented procedures to define specific requirements is inherent throughout the ISO standard. Procedures must be consistent with the requirements of the ISO standard and the stated quality policy. This means that specific requirements for activities being performed will be defined in procedures and/or work instructions. These documents should provide a clear reference or linkage to related procedures or work instructions that define related requirements. For example, a procedure defining requirements for inspecting the product may make reference to another procedure or work instruction to find requirements for handling the product that is found to be out-of-specification.

Food Guidelines (1995) states:

> The company should recognize the balance to be made between documenting procedures and training and experience, and construct the procedures accordingly. The best rule of thumb is that detailed procedures or instructions will be required when the absence of them will adversely affect the level of product quality or service. (p. 5)

"The range and detail of the procedures depend on the complexity of the work, the methods used, and the skills and training needed by personnel involved in carrying out the activity" (ISO 9001:1994, 4.2.2). The standard requires documented procedures "where the absence of such procedures could adversely affect quality" (ISO 9001:1994, 4.9). Note that ISO 9001:2000 has placed less emphasis on the requirements for procedures. Although the choice is being left up to the organization, the importance of defining and communicating requirements in a manner that is useful to the entire quality management system.

The documented structure is usually divided into three tiers: policy, procedures, and work instructions. The policy-level document (the quality manual) addresses the requirements of the standard. Procedures define the activities "who, where, what, why, and when." In some instances they may also address the "how", but typically a complex "how" is defined in work instructions, which may also be referred to as "tier three" documents.

In creating the document hierarchy system, the decision will need to be made as to whether to focus the quality manual entirely on the policy statements or to include both the policy and the procedure-level information together. This decision depends on the system. Often for less complex systems the decision is made to combine the information; however, others find it more useful to keep the information separate. The text of the quality manual must clearly reference or link to the procedure documents for further definition of the requirements, that is, for the "rest of the story." This linkage is most commonly referred to as the documentation "road map" and must be very clear! Do not leave anything to interpretation or assumption. One must know where to go for the defined requirements. The quality manual is only the beginning, the complete documentation structure is the end.

Preparing Procedures and Work Instructions

Compliance with the ISO requirements has the reputation of requiring unreasonable amounts of documentation. However, this does not have to become a monumental task nor is it required to write *War and Peace*-type documents. The following logic applies:

Write down what it is you do.
Do what you say you are going to do.
Document (maintain records) what is done.
Audit (internal quality audit process) to confirm compliance.

It is important to document what must be done, but not to overdocument. If information must be recorded, then the document should state what information is required and where it should be recorded. It does not need to state that "it is required to pick up a pen with the left hand, hold the pen with the right hand and write." Procedures and work instructions should be prepared in a way that is useful to the system. They should define requirements in a consistent manner to ensure that activities are performed in a compliant manner. These documents make an excellent training tool and good everyday reference material.

Many have stated that "writing the procedures was the hardest part of implementation." Associates may be hesitant to define in a written procedure exactly what they do because they are afraid of getting in trouble for not doing it all of the time. These types of issues may be addressed through effective communication and top management support.

Thoughts on Writing Procedures

The hardest part of the system implementation effort was getting everyone to agree on how we do something. It is amazing on how we think we do something the same and then putting it down in a procedure or work instruction identifies so many interesting variations. The most difficult aspect of maintaining the system after certification is remembering when you revise a procedure, that you need to look at all the other procedures or documents that the revision might effect.
—Charlie Stecher, Quality Assurance Manager, Reckitt & Colman, Inc.

One of the most difficult aspects of implementation was ensuring that procedures are always followed and doing what we said we will do.
—Gail Cartwright, Assistant to the AVP Human Resources Department, Bacardi & Company Limited

I would say that the biggest disappointment would have to be the level of documentation and detail required. In many cases, I feel that it would be easy for a company to go a bit overboard with respect to the level of documentation. While the standard does not dictate the level of detail, I feel that some areas are more applicable in different industries and should therefore be more detailed in that business sector.
—Mike Burness, Director of Quality Assurance, Pepperidge Farm Inc.

One of the negative aspects of the system, is people are held accountable to what they write in their procedures. This accountability causes some people not to document their tasks fully. The statements [such as] "It is not part of ISO" or "If I document it, I will have to do it that way" are examples of this problem. We stress to these people that ISO is a tool. We need to use this tool to enhance our system. If tasks are documented correctly, there should be no fear of an audit.
—Bill Lockwood, Package Quality Manager, Hiram Walker & Sons, Ltd.

What has been particularly difficult in our case has been maintaining procedures current. Employees will, in general follow them once trained. Some may not know where in the procedure a particular instruction is, but they know the instruction is there from his or her training.
—Dave Demone, Environmental and Quality Control Manager, & Sylvia Garcia, Environmental and Quality Control Manager, Domino Sugar

The success of our procedures is based on the fact that they were written by the people who do the job.

—Linda Taylor, Training and Quality Manager,
Le Meridien Jamaica Pegasus

One of the best ways we found for documenting procedures was to go right to the source, that is the person who performed the task. We would ask the operator to document what they did. Sometimes, we would perform a brief interview, noting each step that was performed. It also allowed us to note any "missing" steps that were needed but were not being performed to date. We found that the procedures we generated were more accurate, included more of the detail required for the task, and were expressed in a language that was clearly understood by the operator. This was much better than procedures I've seen in the past that were written by someone in the "front office."

—Victor V. Margiotta, Director of Quality, SOBE

The biggest surprise/disappointment was the amount of effort to re-write our existing procedures to fit our new, smaller structure and to conform to ISO nuances. I feel that the hardest part of implementation was writing the procedures.

—Al Gossmann, Quality Assurance Manager, Cultor Food Science

Documentation was one of the hardest parts of system implementation. Having to codify and assemble the quality plans and procedures for the plant was a very large task.

—Henry Gibson, Quality Assurance Manager, Campbell Soup Company

If I could do it all over again with what I know now, I would initially implement only those procedures that directly impact on clauses of the standard and build on these.

—Andy Fowler, Research Engineer/Management
Representative, Bacardi & Company Limited

I would have to say that the greatest benefit achieved was the implementation of a formal Quality Management System. Prior to our certification efforts, the plant did not have much formal documentation. People relied on personal notes and "tribal knowledge" to get things done. As a result, work did not always get performed in a consistent fashion. While there are still some loose ends to work on, the consistency of formal documentation has made an overall tremendous turnaround in the way we do our work.

—Dana Crowley, Production Manager, Danisco Sweeteners, Inc.

One of the most useful benefits that we gained from certification was the consistency achieved through "documentation." We had always been blessed with employees who understood and generally practiced quality concepts, but nearly everything was known by the few "experts," minimal documentation, lots of handwritten scraps of paper in pockets/lockers and each shift did it a bit differently.

—Dave Demone, Environmental and Quality
Control Manager, Domino Sugar

QUALITY SYSTEM 59

Listen to the associates. Create teams that include associates that perform the functions to ensure that activities are documented accurately. These documents are not a "wish list" created by management to define how one might hope an activity is performed. Once "how it is actually done" is recorded, then the teams can work to identify improvement opportunities that may be useful to the process. An example might be that quality assurance would like to circulate cleaning solutions at 175°F, but the system doesn't have the capability, being only able to attain a temperature of 160°F. The procedure should be written using the latter temperature. Improvements in the system can be addressed through the corrective and preventive action processes. If the procedure states 175°F, then records must confirm compliance at that temperature or a noncompliance will exist.

Edward Link (1997) in *An ISO 9000 Pocket Guide for Every Employee* describes the responsibilities for documentation. Management

> must allow those who understand your company the best to work with people who understand the standard very well in an effort [to] create a suitable and effective ISO 9000 compliant quality system. Once that this has been done, they must follow the documented system as all employees should. Reducing variation, ... is a very important aspect of quality. Documenting what we do is the very first opportunity to reduce variation. Once management has enabled the creation of a documented quality system, and it is in place, it is imperative that the system be followed as written. Reducing variation depends on it. It is then the responsibility of every employee to implement the quality system as documented. (pp. 12–13)

Quality Plans

ISO 8402:1994 defines quality planning as

> activities that establish the objectives and requirements for quality and for the application of quality system elements. Quality planning covers product planning, managerial and operational planning, and the preparation of quality plans and the making of provisions for quality improvement.

Quality plans define "how the requirements for quality will be met. Quality planning [must] be consistent with all other requirements of [the] quality system and shall be documented in a format to suit the [system's] method of operation" (ISO 9001:1994, 4.2.3). Quality planning must be defined in a manner such that the outcome of system activities is in compliance with defined requirements. Most systems in the food industry find that the requirements for "how quality will be met" are defined within their documented procedures and reference these as the source of the "quality plan." Many different existing documents such as worksheets, flowcharts, business plans, hazard risk analysis, and product specifications can be used in preparing quality plans for a company's specific products and processes.

The following should be considered when addressing the documentation of the quality plans:

- The identification and acquisition of any controls, processes, equipment (including inspection and test equipment) resources and skills that may be needed to achieve the required quality;
- Ensuring the compatibility of the design, the production process, installation, servicing, inspection and test procedures, and the applicable documentation;
- The updating, as necessary, of quality control, inspection and testing techniques, including the development of new instrumentation;
- The identification of any measurement requirement involving capability that exceeds the known state of the art, in sufficient time for the needed capability to be developed;
- The identification of suitable verification at appropriate stages in the realization of product;
- The clarification of standards of acceptability for all features and requirements, including those which contain a subjective element;
- The identification and preparation of quality records.

—ISO 9001:1994 4.2.3

The definition for "how quality will be met" as related to customer requirements may be addressed in the handling of customer contracts and orders. It is necessary when receiving a customer's order to verify that his or her requirements can be met. The process and requirements to accomplish "how this will be met" should be clearly defined in the related area procedures and work instructions. Some systems that perform design control activities choose to define "how quality will be met" requirements as part of the design development and implementation process.

To summarize, the standard requires

> that [the] company utilize quality planning and consider a stated list of activities in that planning. Quality Planning is all about wanting to hit home runs every time that you are up to bat. Of course, . . . just wildly swinging for the fence will not do the job. Careful planning, analysis of your swing, execution of the determined adjustments to your swing and practice will make the glorious event more likely.
> —*An ISO 9000 Pocket Guide for Every Employee,* Edward P. Link, Quality Pursuit, Inc., 1997, pp. 11–12

Thoughts on Quality Plans

This requires us to document our Quality Management System and to make plans to insure that we make a product that meets our customer's requirements. Quality planning is an integral part of removing waste and doing it right the first

time. This forces us to improve our processes through training and continuous improvement.
—Bill Lockwood, Package Quality Manager, Hiram Walker & Sons, Ltd.

Typically, the food industry looks at quality as something separate from regulatory compliance or prevention based food safety systems. The fact is that quality planning should be an umbrella for integrating quality (quality control points), regulatory requirements (regulatory control points) and food safety systems (critical control points). Food companies should design, validate, implement, verify and update an integrated food safety and quality system.
—Brian Pugliese, Industry Relation Executive, NSF-International

System Nonconformances

Typically when discussing the "quality system," one is referring to the complete system. Note that when auditing to confirm compliance, if there are a combination of several issues reported that demonstrate that the system either isn't effectively established or that the established system's requirements are deteriorating, a serious nonconformance may be raised against the quality system. Such a nonconformance may state that "it could not be clearly demonstrated that a quality system to ensure that product conforms to specified requirements has been established, documented, and is being maintained in compliance to the requirements of the ISO standard."

As a general rule, a noncompliance raised against the quality system is usually serious and means that the system is not effectively implemented, controlled, and/or maintained.

Frequently Identified Nonconformances

- The quality manual does not address all the requirements ["shall" statements] of the standard.
- The quality manual does not make reference to the procedures and/or include an outline of the quality system documentation hierarchy.
- Documentation does not clearly define the requirements for quality planning or requirements are defined but records are not available to demonstrate compliance.
- Procedures are not always available in areas where the activity affects quality.
- It could not always be demonstrated through the interview process that associates are familiar with related procedures and work instructions.

6
MANAGEMENT RESPONSIBILITY (ISO 9001:2000 SECTION 5.0)

6.1 AN OVERVIEW

ISO 9001:2000 puts a greater emphasis on the visibility of "top management" involvement. The terminology has changed from "management with executive responsibility" to "top management." Top management is defined in ISO 9001:2000 (DRAFT) as the "person or group of people who direct and control an organization at the highest level." Auditors will now be looking for more distinct evidence of top management involvement such as presence at meetings, presentations, and communication activities. In the past, in some organizations, "executive management" has technically sat in the background, leaving management review meeting preparation and presentations to the management representative and designated team members. Although this hasn't really been the norm, it does happen in some organizations. The revised standard puts a stronger, more focused emphasis on the involvement of management at the highest level of the organization in setting policies and objectives, making decisions, applying resources, and ensuring that customer requirements are understood, communicated, and met. Although the term "executive management" is used throughout this text, please consider its meaning to correspond with the revised term "top management."

The draft standard Section 5.2 (Customer Focus) states that "top management shall ensure that customer needs and expectations are determined, converted into requirements and fulfilled with the aim of achieving customer satisfaction" (ISO 9001:2000, Section 5.2). This requires proactivity on the part of top management to identify not only individual customer needs but also the needs of the organization's industry through such activities as customer surveys and market evaluations. Of course, the actual performance of these activities will depend on many factors such as the size of the organization and the nature of its product or service.

The draft version goes on to link requirements to ISO 9001:2000, 7.2.1 with the statement that "when determining customer needs and expectations, it is important to consider obligations related to product, including regulatory and legal requirements (see 7.2.1)" (ISO 9001:2000, section 5.2). Although this statement is different from ISO 9001:1994, basically the requirement is similar with more emphasis on defining the methods that the organization uses to stay current on regulatory requirements. ISO's role in requiring that regulatory requirements be met is addressed in detail in Chapter 12.

Requirements for the quality policy and measurable objectives are now defined in ISO 9001:2000, Section 5.3. It has been stressed that objectives must be measurable. The wording of the revised standard may now be more specific in emphasizing that the quality policy be defined in a meaningful manner directly relating to the business objectives. Keep in mind that the policy and measurable objectives must be communicated to associates at all levels of the operation. Records must confirm not only the communication but also the activities, resource assignments, and results in meeting these requirements and goals. The quality policy and measurable objectives (goals) must include the focus on evaluating and ensuring that customer requirements are understood, communicated, progress evaluated, and resources assigned, as required.

"Quality planning" is addressed in ISO 9001:2000, Section 5.4.2, and requires that "top management shall ensure that the resources needed to achieve the quality objectives are identified and planned." Basically, it is specifically required that top management create active plans for change or proactive involvement to ensure that objectives are met. This could include actions from the corrective and preventive action process and any other means that the organization uses to initiate changes and system improvements.

ISO 9001:2000, Section 5.5, addresses the administration, which includes requirements for identifying, defining, and communicating "functions and their interrelations within the organization." This section also defines the responsibilities for the management representative. Requirements for this position have been expanded to include "promoting awareness of customer requirements throughout the organization."

ISO 9001:2000, Section 5.5.4 (Internal Communication) expands on the type of communication through all levels of the operation that must be evident. This particular subject is stressed throughout this text to include awareness and updated training on the quality management system. This may include team meetings, employee/management meetings, letters, posters, newsletters, and whatever means the organization uses to effectively communicate requirements and evaluate performance of the quality management system.

Basically, the quality manual, as discussed in previous sections of this text, must clearly define the organization's policy requirements in meeting the "shall" statements of the ISO standard. Less emphasis is now placed on the requirement for documented procedures; however, this author again stresses the importance of defining requirements to reduce product, process, and system variation. It is again recommended that for those organizations approved to the ISO 9001:1994 standard to reevaluate the current quality manual and defined policy definitions to ensure that the "shall" statements of the revised standard are appropriately addressed. Over time, as the organization and its system complies with the requirements for ISO 9001:2000, evidence (records) must be available to demonstrate compliance.

ISO 9001:2000, Section 5.5.6, addresses the requirements for document control. Basically, these have not changed in that it is still required that the information that associates need to perform their defined responsibilities must be controlled. This revision now only requires that the documents (including external documents) that are required by the quality management system be controlled. Which additional documents may be required will be determined by the organization. It must be cautioned that this does not necessarily mean that all other documents can now be uncontrolled. Do not lose sight of what is best for the system. Remember to integrate the requirements of the ISO standard into the process. Pay particular attention to the discussion on document and data control in Section 6.3 of this chapter, especially the comments on the importance of a sound and effective document control process to the overall growth and maturity of the quality management system.

Requirements for "quality records" are now addressed in ISO 9001:2000, Section 5.5.7, with no major change noted at this point.

The "management review" activity is defined in ISO 9001:2000, Section 5.6.1. Requirements have been expanded to not only include the suitability and effectiveness of the system, but also measurement and discussions of its adequacy. Although the proactivity for management review was implied in ISO 9001:1994, the revised standard stresses the requirement for documented evidence in a more dynamic manner. Although implied previously, the revised standard now specifically states that inputs to the management review meetings must include follow-up to actions identified at previous management review meetings and the identifications, discussions, and assignment of resources to address opportunities for improvement (changes) to the quality management system.

ISO 9001:2000, Section 5.6.3, specifically identifies that the management review output include "actions related to improvement of the quality management system and its processes; improvement of product related to customer requirements; and resource needs." Confirmation that all required management review activities are being performed must be maintained as part of the quality record.

6.2 MANAGEMENT RESPONSIBILITY

The standard requires that "executive" management take ownership of the system. This is done by taking an active role not only in its support and on-hand involvement but also by ensuring that the requirements (quality policy, objectives, customer requirements, etc.) are communicated to associates at all levels of the operation.

The "executive management team" must be defined within the text of the quality manual. This can be one person with responsibility for the entire system such as the president, general manager, or plant manager or it can be a group of individuals such as the ISO steering committee or management staff. This group should include top-level management such as the president/general manager and those positions that directly report to him or her.

Responsibilities for executive management include:

- Define the quality policy and "measurable" objectives.
- Ensure that the quality policy, objectives, and customer requirements are communicated to associates at all levels of the operation.
- Define the responsibility, authority, and interrelationship of personnel who "manage, perform, and verify work affecting quality" (ISO 9001:1994, 4.1.2.1).
- Ensure all necessary resources are provided for the effective operation and maintenance of the quality system. Resources not only include equipment but also the assignment of "qualified" personnel (trained in defined requirements for their responsible activities).
- Assign the management representative.
- Evaluate the system at defined intervals through the management review process to ensure the system's continued "suitability and effectiveness in satisfying the requirements of the [ISO Standard] and the ... stated quality policy and objectives" (ISO 9001:1994, 4.1.3)

Attitude limitations and lack of ongoing support from middle and senior management have been frequently stated as a disappointment. Many times associates hesitate to provide his or her total commitment because of the message they receive from the lack of commitment from management and front-line supervisors. Middle management, in some instances, may present an attitude of skepticism not lending much support. This skepticism of the program is most likely prompted by lack of understanding as to what its structure and discipline mean to their current status within the system.

Thoughts on This Skepticism

One of the most difficult challenges has been the difficulty in getting some middle managers and some operations' department supervisors to realize what an

ISO-based quality system can do for them and how it provides a way of solving their department's problems. This negative attitude is certainly contagious especially to the supervisors's direct reports. There is no doubt that our current system stresses documentation and that it holds people accountable to the quality system much more than pre-ISO. This was somewhat threatening to these people and culture changes slowly, but once it was realized what such a system could do for them and their departments they realized that their jobs actually became easier.

—Tim Sonntag, VP Quality Assurance & Technical, Wixon Fontarome

Successful implementation and long-term effectiveness of a quality management system compliant to ISO 9001's requirements are directly related to management support. The standard emphasizes and demands evidence that this support not only exist but that management is actively involved in system activities. Lack of management support must never be an issue. Compliance activities will become a way of life, from the highest level through all associates of the operation. This is essential to the ultimate success and usefulness of the system. If this importance is not truly believed and communicated by executive management in a manner that is clearly evident throughout the system, then there will be many disappointments and only limited system usefulness experienced. Executive management support and involvement are absolutely essential for the system to grow to its true potential.

Members of middle management in some operations is often only partially supportive because that is what he or she thinks their supervisors want. This type of situation is an executive management challenge, one that must be addressed. Associates at lower levels will perceive this as a type of "lip service," which will directly effect the system's overall effectiveness. Changing the culture, the way one thinks about what is done, following documentation, and being accountable are the most difficult hurdles. However, with management's support and the application of the ISO 9001's structure and discipline, accomplishing this will be one of the greatest rewards resulting in an effective quality management system. Many have stated the biggest adversaries of system compliance become the strongest supporters, once he or she realizes and experiences the system's benefits firsthand.

Thoughts on Management Responsibility

The biggest surprise experienced as a result of certification was the way our company's management came together to become intimately involved in the quality system. There [was] support from the most unlikeliest of places. Some of the positions that I expected to have the most resistance ... provided some of the best assistance. The level of management involvement as a result of certification [was] certainly ... the biggest surprise. Once that level of involvement [was] secured, the drive for improvement became significantly easier.... ISO

certification in reality simply provided the framework for [managers] to solve their own problems.
—Russ Marchiando, Quality Systems Coordinator, Wixon Fontarome

One of the hardest parts of our implementation effort was in getting managers and their respective staff to take responsibility for the process and to be committed to ensuring that the quality system is effective.
—Anonymous

Management involvement is crucial to success of the system. While certification itself is a rewarding goal, it should not be the reason for undertaking the ISO endeavor. Management involvement can be garnered by developing a strategic plan for system implementation as it relates to the business. An example of this could be the inclusion of economic quality indicators (e.g. cost of product manufactured/sold, EBIT, consumer satisfaction indices, etc.) in management review. Relating the executional effectiveness of the system to these indices easily demonstrates the impact of a rigorous quality system to upper management. Continuously relating the elements of ISO to standard business indices facilitates support and more importantly, commitment to the system.
—Mike Burness, Director of Quality Assurance, Pepperidge Farms, Inc.

Executive management on-hands involvement was seriously missing during the initial implementation activities. This overall was detrimental to a timely and effective implementation; however, once management "saw the light" the process took off.
—Anonymous

Quality Policy and Measurable Objectives

The quality policy is a statement for the commitment to quality. "The quality policy [should] be relevant to the organizational goals and the expectations and needs of its customers" (ISO 9001:1994: 4.1.1). It can be very simple, only a few words, or it can be complex taking two or more paragraphs.

Thoughts on the Quality Policy

[The quality policy] states what we want to accomplish for our products, people, and processes. [It] should be measurable. . . . If you cannot measure it, how do you know if it is working? This keeps us [focused] on what we say in the Quality Policy. It should also be kept simple.
—Bill Lockwood, Package Quality Manager, Hiram Walker & Sons, Ltd.

The quality policy statement should be kept as simple as possible, while clearly addressing the full scope of the business. Examples of some quality policy statements may be as follows:

- The ABC Juice Company will manufacture the best possible product and will comply with customer requirements and internal specifications.

- The Exotic Flour Company is committed to quality and will manufacture a safe, wholesome, product that conforms to all specifications and is delivered to the customer on time.

Some companies combine their mission statement with goal statements as part of their quality policy and stated objectives. This is beneficial because it ensures the system involves the entire business scheme.

Objectives must be measurable and should be defined at the time the goal is identified. They should be related to the organization's own activities, important to meeting defined system requirements and the needs of the customer. Measurable objectives are used to evaluate the overall suitability and effectiveness of the system and status in achieving the quality policy. An example would be to state that "driving habits were going to be changed to improve mileage." To ensure that efforts are effective, results must be "measured." Thus one may record and track the mileage and total gallons used to measure the improvements. Another example would be to state "eating habits were going to be improved to lose weight." Progress may be measured by monitoring pounds or clothes size.

The quality policy may state "to provide the best possible product" with an effective measure being the "percentage of product meeting specification" or some other internally defined means for measuring product acceptability.

Let's emphasize that the quality policy must be communicated to associates at all levels of the operation. Be sure that posted copies are maintained as controlled documents. This would mean the individual posted copies are identified in a manner such that if the policy is revised, then each copy can be retrieved and replaced with the most current version. In addition, the policy should be signed by the current president or general manager.

Thoughts on Measurable Objectives

Russ Marchiando (Quality Systems Coordinator, Wixon Fontarome) stated:

> As management became increasingly involved it became important to measure the system's performance to identify areas where we could improve and also to validate the return on the company's investment in the quality system. We examined what would be the most important quality related issues to our customers and found that product integrity, functionality, quality, on-time delivery, order accuracy and plant sanitation were paramount. We also looked at what was important to our company from within and found that meeting budget goals, increasing inventory turns, reducing quality control holds and observing proper lead times were the most important issues to maintain our business system. It was then determined that these key areas could be put into four categories:
>
> *Quality Goals*—QC holds and plant sanitation
> *Customer Satisfaction Goals*—reduction of internal/customer complaints and a reduction of customer returns

Efficiency Goals—on time delivery, order accuracy and proper lead times
Return on Investment Goals—meeting budgetary goals and increasing inventory turns.

These key areas are now measured each quarter. Predefined goals have been established for some areas while other areas are still in a data collection process to establish baseline measurements. Our priority focus is to improve on the numbers each time they are measured. This relates to achieving the quality policy while measuring the suitability and effectiveness through these defined objectives or goals.

It is important that the ISO-compliant quality management system be integrated into the company's structure. It is difficult to gain full benefit of such a system if it is kept separately. Statements such as "that is ISO" and "this is non-ISO" result in confusion and weaken the overall process.

Communicating the Quality Policy

The standard requires that the quality policy and objectives be clearly communicated to associates at all levels of the operation, promoting full understanding of the importance and role of the ISO-compliant system throughout the organization. Auditors will confirm this through the interview process, asking associates about the ISO system and the quality policy. An auditor should not expect an associate to necessarily have the policy memorized, however, the associate should understand the role he or she plays within the system to contribute to accomplishing these goals. A good answer when questioned on this role would be that "my role is to do the best I can to ensure that the product meets the requirements." An auditor will be able to assess the associate's knowledge and enthusiasm regarding his or her concept. Also associates should know where to find a controlled version of the policy (i.e., posted on the wall, in an area manual, etc.).

Management Representative

The standard requires that

> management with executive responsibility . . . appoint a member of [its] . . . own management who, irrespective of other responsibilities, [will] . . . have defined authority for ensuring that a quality system is established, implemented, and maintained in accordance with [the ISO Standard and for] reporting on the performance of the quality system to management for review and as a basis for improvement of the quality system.
> —ISO 9001:1994 4.1.2.3

This means a person with authority is assigned the responsibility for overseeing implementation and maintenance of the compliant system. This assigned authority is important in order to put the importance of this responsibility in its proper perspective. Many times management and associates assume that the system belongs to the position of management representative. That it is the responsibility of that position to ensure that it succeed. System implementation, control, and maintenance must be a team effort that is evident throughout the entire system.

The management representative must have support from not only the executive management team but also from supervisors and associates throughout the system. It is not the management representative's system, it is everyone's system and its success is directly related to the best efforts of the entire team.

Let's consider the analogy of a pitcher, who can't pitch a no-hitter without the full roster of players doing the best they can at what they are best at. The executive manager or management team would be synonymous with the owner of a baseball team such as George Steinbrenner. The management representative would be the team manager or the head coach. The owner would provide the resources and backing to trade for the best players; then the manager directs these skills in an effective manner to establish, implement, and maintain the goals of the team. The goals could be "measured" through win / loss records, batting averages, on-base percentages, and the like.

Management Review

Effective management review meetings are essential to the long-term health of the quality management system. The ISO standard states that these evaluations must be held at defined intervals sufficient to ensure the system's suitability and effectiveness. Although the ISO standard doesn't require a specific frequency, most define it to be no less than every 6 months and possibly even quarterly for an immature system. The frequency must be frequent enough to be able to monitor the effectiveness of the system. However, evaluations should not be so frequent that evidence provides inconclusive information for measurement of the suitability and effectiveness of the system. In other words, several months of data may be necessary to evaluate a measurable objective.

Thoughts on the Management Review Process

> The management review process can very quickly become the sole responsibility of the overzealous management representative and care should be taken to avoid this situation early in the implementation process. While the management representative is required to report on the quality system to the meeting, the meeting would be best chaired by a member of executive management. This will allow for ownership of the quality system review process by upper management (executive management). Their involvement is crucially important to the success

of the system. Another negative is the unfortunate inability to score the "early wins" in the management review process. A struggling system may not always show the early success that cost conscious managers are looking for to justify the system's implementation. It is important for all to realize that the system will only be as good as the company as a whole will allow it to be. Culture change issues, a lack of genuine management support and other conflicting variables will undoubtedly provide uncomfortable bumps in the road early on. The management review process under a management team that is genuinely committed to the quality system provides a voice for quality that few organizations may have. It is the driving force for quality improvement and provides peer accountability for the completion of quality objectives. No one wants to attend the next management review meeting without finishing all of the items that were assigned to him or her at the last. The management review meeting should also be used by the management representative to acquire help in the implementation effort if it is needed. The management representative has the ear of the company's management and should take full advantage of the opportunities available to them.

—Russ Marchiando, Quality Systems Coordinator, Wixon Fontarome.

Numerous management changes due to an acquisition required a solid mature system to help our organization through this time. Because our quality management system was founded more on the philosophy of certification rather than business improvement, we struggled when the business environment became more competitive. As one that is involved in overall business improvement, I am amazed at the ability of the ISO management review meetings to measure with reasonable accuracy the state of our business. When our quality management system was struggling, there were numerous other areas of our business that were struggling.

—Anonymous

Management review agenda items should include at a minimum the following:

- Review of the quality policy and its continued application to the defined system.
- Review status in achieving measurable objectives.
  - Assign resources as appropriate to aid in achieving these objectives.
  - Restate objectives, as appropriate.
- Review internal quality audit process.
  - Results of audits.
  - Audit schedule, on-time performance of audits.
  - Timely and effective response to corrective actions identified during the audits.
  - Follow-up to confirm effectiveness of actions taken.
- Review of corrective and preventive action process.

- Confirm process includes product, process, and system issues.
- Timely response.
- On-time completion.
- Follow-up to confirm effectiveness of actions taken.
- Discuss specific preventive action type activities.
  - To ensure "appropriate sources of information" are being reviewed to "detect, analyze, and eliminate potential noncompliances" (ISO 9001:1994 Section 4.14.3a)
- Discuss customer requirements.
  - Ensure requirements are understood and communicated, as appropriate.
  - Discuss any issues and concerns in meeting customer requirements.
  - Review customer complaints and address, as appropriate.
- Review status of action items from previous management review meetings.
- Summarize action items to be addressed and responsibilities.
  - Identify areas requiring progress updates at subsequent management review meetings.
- State a conclusion on status and effectiveness of the system based on items reviewed at the meeting.

These should be defined in the related, procedures or work instructions, providing guidance and consistency for the management review meeting process.

More Thoughts on the Management Review Process

If you do not measure and review your results, you cannot tell how well you are doing.
—Bill Lockwood, Package Quality Manager, Hiram Walker & Sons, Ltd.

One of the most useful benefits from the ISO compliant system is the management review. It provides a vehicle for top management and key players to review, evaluate and improve the system in a disciplined and objective manner on a routine basis.
—Eric Halvorsen, Quality Assurance Manager—Auditing, Campbell Soup Co.

Quality Records

Records of the management review meetings must be maintained. One of the most effective means to accomplish this is to maintain an agenda and minutes summarizing the complete discussions. Presentations, charts, graphs, and other material distributed during the meeting to substantiate the discussions should be identified (attachment A, B, etc.) and then referenced in the meeting minutes.

DOCUMENT AND DATA CONTROL 73

Frequently Identified Nonconformances

- It cannot be clearly demonstrated that executive management has defined the quality policy and objectives related to the system's commitment to quality.
- Through the interview process, it could not be clearly demonstrated that the quality policy has been communicated to associates throughout all levels of the operation.
- It is evident by the lack of compliance status in various system areas that resources have not been adequately provided to the system.
- The management representative does not have sufficient responsibility and authority to ensure that the quality system is compliant.
- Management review meetings are not being performed at the frequencies defined in related documents.
- Management review meetings are being held, but meeting minutes do not provide evidence that one or more of the requirements (e.g., review of the results of internal audits, corrective and preventive action activities, etc.) are being evaluated at these meetings.
- Records of management review meetings are not being maintained, not identified as quality records, and/or not complete. Referenced review items are not being maintained as part of the record.
- The posted quality policy is not signed by the current general manager. Note that the current general manager has been in his or her position for 6 months; however, the quality policy still represents the support of the previous manager rather than the current one.
- Although implied, minutes from the management review meetings did not include a statement on the effectiveness and suitability of the system as a result of the items reviewed.

6.3 DOCUMENT AND DATA CONTROL

The requirements for controlling documentation focuses on ensuring that all documentation and data used within (related to) the quality management system is "controlled." There must be a method of controlling system documents (policy manuals, procedures, work instructions, etc.) and data (specifications, process parameters, etc.) so that pertinent issues of documents are available where needed and that obsolete documents are promptly removed.

The policy manuals, procedures, and work instructions must, of course, be controlled. Other documentation such as specifications for raw materials and final product, process parameters, artwork for packaging, current regulations, and codes of practice which may be utilized, must also be controlled.

The process must include a means to ensure that documentation is available to all areas responsible for the activity. For example, if a moisture test is being performed at three different locations within the process, then a current document (i.e., work instruction) that defines the requirements for performing the moisture test should be readily available to each area.

Always keep in mind that as documents are reviewed and revised, extra care should be taken to ensure that the "road map" through the documentation to related work instructions, procedures, and the like is clearly defined.

Procedures and/or work instructions should define responsibility requirements for document control activities, revision protocol, distribution of the revised documents, and destroying the obsolete versions. In some instances obsolete documents may need to be maintained for specific reasons such as for historical data, reference, and regulatory issues. In these instances, the document must be appropriately identified as obsolete.

In the world of processing, it is very common that different versions of the same activities, memos, documents, and the like may exist. Document control focuses on ensuring that everyone is working off the same version of a document. The initial phases of system implementation and finding and collecting all this material can be an "exciting" challenge.

Edward Link (1997) in *An ISO 9000 Pocket Guide for Every Employee* provides an excellent description of the importance of document and data control to the entire quality management system:

> The importance of this element to your quality system should not be understated. It is the greatest source of nonconformities for quality system audits. Think in terms of your own company and how many times the paperwork has been the source of the error. When the documents are unavailable, out-of-date or inappropriately changed, they can cause elevated scrap and rework. (p. 20)

Thoughts on the Importance and Role of Document Control on the System

> Document and Data Control is one of the primary benefits resulting from ISO registration. It instills a discipline and accountability that many companies lack. This discipline promotes improved and effective performance, traceability and troubleshooting.
> —Jim Murphy, Manager of Design Process and Validation, The Dannon Company

> Our processes are based upon procedures and work instructions. These documents need to be kept up- to- date, in control and used consistently. If we do not comply with these simple words, we do not control our processes, they control us.
> —Dana Crowley, Production Manager, Danisco Sweeteners, Inc.

DOCUMENT AND DATA CONTROL 75

External Documents

Document control not only includes internal procedures and work instructions created for the process but also external documents. External documents are those documents that are required by the system but not generated within the system. Examples of external documents would include some of the following:

- ISO 9001:2000: If it is stated that the system is compliant with the current version of the ISO 9001 standard, then the appropriate version of the standard must be available within the system.
- AOAC testing procedures: If a specific test is required to be performed according to an AOAC method, then a controlled copy of that test method must be maintained and available.
- Manufacturer's equipment manuals: If the system defines that it is required to perform an activity "according to manufacturer's requirements," then where those requirements are defined (i.e., manufacturing equipment manuals) must be maintained as externally controlled documents.
- Customer specifications and drawings.
- Regulatory laws and guidelines such as GMPs (Good Manufacturing Practices), HACCP (Hazard Analysis Critical Control Point) Codex Alimentarius, weight charts, and the like.
- Standards such as color and/or label standards.
- Corporate manuals and training documents.
- Supplier contracts used to define specified requirements. These are commonly used with pest control suppliers.

A master list of the external documents must be maintained, including the identifying factor for the document, its distribution and location, and the responsible source to ensure that these documents are updated, as appropriate. The documents themselves should be identified as controlled. It is recommended that the documents are also individually numbered to keep up with the number of controlled copies maintained within the organization.

It is recommended, whenever useful and appropriate, that requirements stated in the external document be incorporated into a system procedure or work instruction. For example, if the manufacturer's requirement for calibration is twice per year and that is deemed acceptable to your system, then state this in the procedure or work instruction for calibration. If the procedure or work instruction states that calibration will be done according to the manufacturer's requirements, then the documents that define the manufacturer's requirements must be controlled as an external document. This can become burdensome and impractical. It would also mean that the associate would have to look up the requirement every time, when it would be easier just to read it in the internal procedure or work instruction. Also in some instances, the manufacturer's recommendation may not be appropriate for your system, being either too stringent or not stringent enough. If, however, there is a multitude of information required, such as in an AOAC testing manual, it may be more advantageous to control that document rather than to include the information into a system procedure or work instruction. Keep in mind though, if the manual contains a hundred tests and your system only needs one or two, it might be better to include the tests in work instructions. Remember it is your system to define. When making these types of decisions be sure to be cognizant of your customer's requirements. A customer requirement may specifically identify a testing procedure. In that instance, a copy of that specific test procedure should be maintained as an externally controlled document. Keep in mind that your system may not always require you to purchase a new volume of an external document if the volume referenced and used contains the proper information. This depends on the particular external document and the manner in which the system defines the requirements. For example, the testing book *Standard Methods for Testing* may be revised each year, but the particular test that is being referenced and used within the system does not change; thus, there should not be a requirement to have to purchase an expensive new volume each year. Another appropriate example is the recent revision to the ISO 9000 standard. If your system is currently approved to ISO 9001:1994, then a controlled copy of that standard would be required by the system. The ISO 9001:2000 would not apply until the time the system is approved to that standard.

Reference Versus Controlled Documents

As external documents are identified, remember that these are documents required by the system but not created within the system. A reference document may be a document referenced for information but not specifically required by the system. It may be advantageous to stamp these documents as "reference" to clarify their place in the system.

Forms

Identifying and managing forms used in the system can be very confusing. A blank form would be considered a controlled document because it contains required current information. Once the form is completed with data documented confirming compliance to a specified requirement then it would be considered a quality record. Keep in mind that depending on the information on a particular completed form and requirements defined within the system, not all completed forms may be considered a quality record.

Forms can be controlled with a date, revision number, or any means that will make it possible to confirm that the current revision is being used. A helpful tip with forms is to also reference the identification of the work instruction or procedure that defines the requirements for the specific forms. This provides an excellent guide for associates should they have a question on the use of the form and want a quick reference where to go.

Also, keep in mind that a form demonstrates compliance to a specified requirement. Be sure that the form is addressed somewhere in the system. Many times forms and information will be identified as a quality record, but it is unclear exactly what the information is actually demonstrating compliance for. More will be discussed on the specifics for forms as Quality Records in Section 6.4 of this text.

Posted Documents

During the implementation phase, identifying controlled documents can be a challenge. Most likely many documents have been posted throughout the operation over the years. This is the time to review this information and identify what is required by the system. If the document is providing information important to requirements of the process, then it must be controlled. Remove these documents from posting and include the information in a procedure or work instruction. Once contained in a controlled document, if appropriate, it can be reposted. Keep in mind that procedures and/or work instructions, even a particular page, can be posted as long as the master list identifies the document (page) and its location. A rule of thumb for determining if the information should be contained in a controlled document would be to ask that "if removed, would key information for and about the process be missed?" If the answer is yes, the information should be in a controlled document.

Some systems actually identify the use of "temporary" documents, which provide information on temporary changes, trial activities, and the like. However, be careful, these types of situations should be monitored closely to ensure continual compliance. Posted documents that provide information should be identified as "for information only." It is recommended that these be dated and removed after a specified time period.

Edward Link (1997) in *An ISO 9000 Pocket Guide for Every Employee* provides the following guidance to employees on the use of controlled documents:

> Optimizing document control is extremely important to the reduction of variation. The documents that are provided for your use will carry some mechanism indicating that they are controlled documents. . . . Be sure that you have controlled documents before you use them. Remove any document from your area that is not controlled in accordance with the applicable procedure for that document type. This includes all quality system documents conveniently taped to walls or equipment in your workstation. You may replace them with controlled versions as long as the document control system has provisions that assure removal and replacement at the time of change. If the information in those documents is required, take the steps necessary to get it into a controlled document. (p. 24)

The Master List

It is necessary to maintain a master list or an equivalent that identifies the current revision status and distribution location for all controlled documents. This is important in protecting against the inadvertent use of obsolete or invalid documents.

Some systems maintain one master list controlled by a document control coordinator. Others choose to have the different areas maintain their own master lists. In this instance, the department coordinator would maintain the identification and distribution of the controlled documents for which he or she is responsible. The management representative or his or her designee would then maintain the master list of the quality system documentation. Either way can be effective. It is your system to define.

One word of caution, when there is one person controlling all the documents, it can get burdensome, especially if that person is also charged with keeping the area books up to date. It has been found that to achieve a higher level of effectiveness, the document coordinator should issue the area documents to the department manager or his or her designee. The manager should then take responsibility for updating the manuals within his or her department. In doing so, this will give the manager the opportunity to review the changes and confirm that all required training on the updates is performed.

Approval and Issuance

This is a straightforward requirement, which basically states that "documents and data shall be reviewed and approved for adequacy by authorized personnel prior to issue" (ISO 9001:1994, Section 4.5.2).

Handwritten Changes

Generally most systems do not allow handwritten changes to a controlled document. It is required that the changes be made on an uncontrolled copy of the document, then submitted for approval and subsequent issuance. However, some systems do allow handwritten changes as long as they are completed by individuals identified as having the proper authority. Care must be made in this situation. All copies of the same controlled document must be identical. Thus, if a handwritten change is made on one controlled copy, be absolutely sure that this same change is made on every other controlled copy of that document. This can be a tough situation to maintain over time, but some do accomplish this successfully. It requires strong focus and diligence to be sure it doesn't slip.

Identifying the Nature of the Change

"Where practical, the nature of the change shall be identified in the document or the appropriate attachments." (ISO 9001:1994, Section 4.5.3). This can be accomplished in many ways. The two most popular are to highlight the revisions or describe the revision in a revision history section of the document. The revision history identifies the section revised and briefly describes the nature of the revision. This information can be very useful to the document user. The system is yours to decide; however, as with all requirements, be sure that they are clearly defined in a procedure or work instruction.

Training on the Updates

The system should define the requirements for training on document revisions. An effective means to accomplish this is for the document itself to identify whether training is required. Often procedures and work instructions will be revised because of typos, wording changes, or even format changes. The distinction as to whether or not training is required helps alleviate some of the burden that could develop if your system required training on every revised document no matter what the revision. Should training be required, then a quality record to confirm that this has been performed must be maintained. Some systems actually have the associate sign the first page of the document to confirm that he or she has reviewed the change and understood it. It is again your choice as long as the activity is being performed with a record available to demonstrate compliance.

Don't Forget to Control the Data

Document control requirements must also be applied to "data" control. Software programs that are used to maintain controlled documents, records, or other system activities must themselves be controlled. For the most part, control of software programs is usually defined as software backup activities supported by a log or other defined record maintained to demonstrate compliance.

A Controlled Document or a Quality Record

Many times confusion develops in understanding the differences between a controlled document and a quality record. Document control is a process to "control" documents to ensure that the most current versions of instructions, requirements, and the like are available to those that must have the information to perform required activities. A quality record is a record or proof that is maintained to demonstrate compliance to defined requirements. A blank form may be a controlled document to ensure that the current version is in use. Once information is recorded on the form, it becomes a record proving that an activity was performed. In some instances, a document may actually be both a quality record and a controlled document. An example would be the approved supplier list. The responsible person for maintaining information regarding approved suppliers would maintain this list as a quality record to demonstrate compliance to defined requirements. Departments that order from approved suppliers must have access to the current listing. The list would then be distributed as a controlled document to those specific areas of responsibility. Related procedures and/or work instructions would define distribution requirements.

The Controlled Identification

There are many different methods used to identify a controlled document. The most popular method is to print it on specially designed stationary that is used only for that purpose. Some stationary is designed to change to an off color when copied. Another popular method is to have a colored stamp used to mark "Controlled" on each page. Many times the stamp itself states something to the fact that "the red color identifies the document as controlled." Whichever method is chosen, be sure requirements are clearly defined in related procedures or work instructions. Associates must be trained on these requirements with records maintained to demonstrate compliance. Also ensure that the controlled paper or stamp is kept in a secure location with access only to those individuals having the authority to issue the controlled documents.

Software Programs

There are many different software programs available for document management and document control. Through experience, the true effectiveness of any software program depends on the users and the system to which it is being applied. Contact the various suppliers and request demonstrations and other appropriate literature. Clearly define what you expect the program to do for your system. Communicate these requirements to the potential suppliers. Ask suppliers for the names of some of the companies that use their software and contact them for their comments and recommendations. The importance of talking to users can not be overstated. Always ask about the support and response time regarding technical questions and requests for assistance. Technical support after purchase is a must! Some operations manage document control very successfully by using such Microsoft programs as Word or Access. It will be your choice. Keep in mind though that the quality of training and technical support provided by the software company could be proportionate to your operation's success in using its software program.

The Paperless System

More and more systems are going to a "paperless" system, maintaining controlled documents on a database. This can be very effective; however, ensure that associates truly understand and can access their related procedures and work instructions. There may still be a need for printing documents for training, auditing, and the like. It is recommended that a footer be added that automatically identifies the print date and makes a statement such as "this document expires five days from the print date." With that sort of statement, there will be no question as to the controlled verses uncontrolled status of a document.

Remember that software programs used for controlling information and for maintenance of records must be controlled to ensure their content is current and accessible. Controlled documents accessible through the database must be "read only" access to the users. Access for revisions must be controlled ("password protected") and only available to those associates who have the defined authority to make changes to documents. How this is managed and the location of the controlled documents (e.g., "G" drive), should be clearly communicated and defined in appropriate procedures or work instructions.

More Thoughts on Document and Data Control

> Having to maintain (revise) the quality system documentation is the biggest challenge.
> —Henry Gibson, Quality Assurance Manager, Campbell Soup Company

One of the most difficult aspects of maintaining the system after certification has been

> [maintaining] document and data control revisions. Focus must continue to ensure that system documentation is truly beneficial to all that use it and does not become overbearing. It is crucial that the ISO system be used daily and that all employees experience the value of the system as a foundation to everyday operations and thought processes.
> —Jim Murphy, Manager of Design Process and Validation, The Dannon Company

Robert Peach (1997) in *ISO 9000 Handbook, Third Edition*, provides the following quote:

> [David] Middleton pointed out that document control is a common weak link found during third-party audits. He said this problem is usually remedied by identifying "relevant documents" (those that directly impact product or service quality) and by demonstrating document control by date, review status, approval, and master list. (p. 93)

Edward Link (1997) in *An ISO 9000 Pocket Guide for Every Employee* provides the following insight on management's role:

> When management understands the contribution that poor document control makes to losses, they quickly focus on the implementation and maintenance of strong document control systems. During their management review sessions they

should be especially attentive to the document control related issues that are reported through internal audits or surfacing as root causes for problems preventing the attainment of intended goals. (pp. 23–24)

Frequently Identified Nonconformances

- Numerous copies of uncontrolled documents were observed posted in various work areas throughout the process.
- Related procedures and work instructions do not define the requirements for maintaining and controlling the document control software system.
- Related procedures and work instructions do not define the requirements for the control of forms to ensure that the most current version is in use.
- Although it was stated that procedures and work instructions are maintained on the database, associates could not demonstrate familiarity with accessing the related documents in their areas.
- In several areas, posted controlled documents were not the same version as those identified on the master list.
- It was noted that several controlled documents had handwritten changes, which are not permitted according to work instruction DC-01 Rev. 2."
- Although work instruction DC-42 Rev. 0 states that handwritten changes are allowed by the person authorizing the document, in several instances, the handwritten changes on the area document did not correspond with the content of the master copy of the same document.
- System documentation states that a master list of documents would be maintained in each department; however, several departments could not either present a master list or the master list was outdated.
- Although a master list of documents was being maintained in each department as required by document control procedure DC-01 Rev. 2, this list did not identify the revision status and/or the distribution of these documents.
- Several documents such as the Federal Register CFR 21 and equipment manuals for the refractometer were referenced in area documents; however, evidence was not available to confirm that these documents were being controlled as external documents.
- There were several examples noted in the process areas of an obsolete version of form PCF-02 Rev 2 being used.

6.4 CONTROL OF QUALITY RECORDS

Procedures and/or work instructions must define the requirements for handling, maintaining, and disposing of records that are used "to demonstrate conformance to specified requirements and the effective operation of the quality system" (ISO 9001:1994 Section 4.16). "Quality records are those which demonstrate the effective management of the system and should therefore be retained" (*Food Guidelines*, 1995, p. 18). This includes specific records as identified within the actual ISO standard. The ISO 9001:1994 standard identifies those required system records with a reference to 4.16 in parenthesis whereas the ISO 9001:2000 draft references section 5.5.7 in parenthesis.

The following also applies for quality records:

- Identify retention times and responsibilities and requirements for identifying, maintaining, filing, and disposition.
- Storage must be in a manner that protects the records from loss, damage, and deterioration.
- Accessibility within the system, as appropriate.
- Legible and of a permanent nature.
  - Should not be completed in pencil or corrected with "correction fluid."
  - Never erase.
  - Corrections/changes should be initialed and dated by an approved authority.
  - Explain blank spaces (i.e., process not running, etc.).
- Make accessible to customers as required by the specific customer's contract.
- Be of whatever means appropriate to the system (i.e., hard copy, electronic data, etc.).

It is not necessary to save every notation. It is necessary for the system to identify which records and how long they must be maintained to demonstrate compliance to a defined requirement. One company I worked with did many tests on finished product. Although associates recorded the work on various logs, the actual quality record maintained to demonstrate compliance was called the Primary Daily Laboratory Log Sheet, which at the end of the day, included the results of all tests performed.

Food Guidelines (1995) provides the following thoughts on quality records:

> All systems generate records. Some records are more vital than others, some need to be kept for a long time and may be subsequently required to prove your product met the specific action when you sold it. An effective record control system is vital. It is important that records are correctly identified and stored in a manner which makes them retrievable, whether they are as hard copy or on electronic media. It is also important that the system does not become cluttered with unnecessary old records. Keeping everything forever is not an answer to records management. The system must define what to keep, who keeps it and for how long. When records are to be destroyed, there should also be instructions for doing this. (p. 39)

The standard does not specifically state how long records should be kept. This must be an internal decision based on the duration required to demonstrate compliance to specified requirements. For example, depending on the record, 3 to 6 months of daily records may be sufficient to demonstrate compliance; however, monthly records may have to be kept for at least 12 months. Many registrars require that system maintenance items such as internal audit reports, management reviews, customer complaints, corrective actions, and preventive actions be maintained for, at a minimum, 3 years. This is usually the time period required to demonstrate that the quality system is being maintained in an effective manner. It also generally reflects the approval period or life of the ISO-compliant certificate. These records are important sources of information when evaluating the overall effectiveness of the system and adequately planning for the recertification or triennial assessment. Further consideration when defining the retention times may include regulatory requirements, company legal policies, shelf life of the product, and customer requirements.

Quality Record Data

The data recorded on the quality record should be legible and able to be clearly understood. Be sure that records contain all pertinent information. Recording the required data is in response to directions outlined in controlled documents (procedures and work instructions). Keep in mind that it is important to clearly define what the required data will be to demonstrate "conformance to specified requirements and the effective operation of the quality system" (ISO 9001:1994 4.16).

Controlled Documents Versus a Quality Record

Many times confusion develops in understanding the differences between a controlled document and a quality record. Although this was discussed in regard to document and data control, it is worth the emphasis in relation to quality records. Document control is the process to control documents to ensure that the most current versions of instructions, requirements, and the like are available to those that must have the information to perform required activities. A quality record is a record or proof that is maintained to demonstrate conformance to defined requirements. A blank form may be a controlled document to ensure the current version is in use; but, once information is recorded on that form, it becomes a record proving that an activity was performed. In some instances, a document may actually be both a quality record and a controlled document. An example would be a production schedule. The position responsible for creating the production schedule would maintain these as a quality record to demonstrate compliance to the defined requirements. Since the schedule includes information required for specific activities, it may be distributed as a controlled document to specifically identified areas of responsibilities. Related procedures and work instructions would define requirements for distribution, revisions, and identification of the current version.

Maintaining a Master List of Quality Records

There are several possible means of identifying the quality records. The most common is to maintain a master list. Procedure, work instructions, and other controlled documents reference this master list for the defined requirements for individual quality records such as responsibility, storage location, and retention times. All associates responsible for a quality record must be trained in the quality record procedure and work instruction. If the master list defines requirements for a record or records for a specific responsibility, then to perform this responsibility the associate must have ready access to the master list.

Some systems identify each record and all requirements for retaining that record in specific area or departmental procedures and/or work instructions. In some instances, different areas may have the same record in common. A specific requirement such as the retention time may be revised in one area's documentation while being overlooked in another. This will result in conflicting requirements. Over time, as the system matures, this type of situation will occur more frequently with the system very likely reaching a major noncompliant situation. It may be easier to maintain and avoid in smaller systems, but still in time will become a system burden and increase the likelihood of creating a major noncompliant situation. The advantage of having one list is that there is no doubt where the requirements are defined, and should a revision be required, then it can be completed in a simple and effective manner.

Electronically Controlled Records

Quality records may be maintained electronically. However, as discussed in regard to document control, the process for controlling the software databases (i.e., backup) to ensure that they are protected and contain the appropriate information must be defined and supported by records demonstrating compliance.

Food Guidelines (1995) describes electronic data as follows:

> Many records may be stored for their active life as electronic data in computers. Consideration must be given to maintenance of this data both by "back-up" arrangements whilst live and for long term storage. It is not necessary to keep records as hard copy if a suitable alternative is available (e.g. optical discs, magnetic tape, etc.). (pp. 39–40)

Thoughts on Quality Records

The most difficult aspect of maintaining the system after certification is the maintenance of good records and attitudes especially when going through periods of constant change to processing technology. A systems coordinator is a very important person to have in place to ensure compliance in all respects.

—Rex. N. Gadsby, retired GM and CEO Dairy Industries, (Jamaica) Ltd./Grace Food Processors, Ltd.

Frequently Identified Nonconformances

- Conflicts exist between the actual title and name of a quality record and its listing as a quality record. (In other words, the title or identifying factor on the record does not match the identifying factor on the list.)
- Records identified on the master list are not referenced in related area documents such that it is unclear exactly what functions or activities for which the records demonstrate compliance.
- Associates were not familiar with the defined requirements for quality records as related to their areas of responsibility nor did they have ready access to the master quality record listing that identified records for which they were responsible.
- Although it was stated that records were being maintained for 3 years, records from quality audits performed 2 years prior (system had been functioning for 4 years) could not be found.
- Review of a sampling of completed quality records showed many blank spaces with no explanation of why the data was not being documented. Note that it was stated that the operation was not operating; however, the records did not identify this.
- Procedures did not always identify or make reference to required quality records.
- Review of a sampling of quality records showed evidence of changes (cross-outs, erasures, etc.) with no identification or approval for these changes as required by procedure DC-02 Rev 2.
- Procedure TR-01 Rev 2 identifies the retention time for training records as ongoing. Although it was stated verbally that this translates to "forever," this was not clearly defined in the procedure.

7
RESOURCE MANAGEMENT (ISO 9001:2000 SECTION 6.0)

7.1 AN OVERVIEW

This section of the revised standard begins with the defined requirement for the "provision of resources." It is very specific stating that "the organization shall determine and provide, in a timely manner, the resources needed to implement and improve the processes of the quality management system, and to address customer satisfaction" (ISO 9001:2000 Section 6.1). In other words, evidence must be available that the system has the resources it needs to not only be effective but to also apply improvements and to ensure customer satisfaction.

Training requirements and the definition for "qualification" is defined in ISO 9001:2000 Section 6.2. These requirements have been expanded to place a greater emphasis on the evaluation of competency, effectiveness of training, and ensuring that associates are aware of the "relevance and importance of their activities and how they contribute to the achievement of the quality objectives" (ISO 9001:2000 Section 6.2 d). It will now not be sufficient to just perform training; follow-up to confirm its effectiveness will be required. Training records must include documentation of both the positive and negative results of training. Although, at first, organizations may cringe at this requirement, the logic and potential for the overall positive impact that this should have on the business of the organization will be more than worth the extra effort. Most successful organizations have already identified the merits of this activity and are performing this type of evaluation to some degree.

The requirement that associates understand and be aware of the role they play in meeting the policy and goals of the organization is another aspect that most organizations are already performing. This generally has occurred through the communication of the quality policy, what it means to the company, and the role the associates from all levels of the operation play in achieving it. Efforts to continue this focus must now be approached in a more aggressive manner with records maintained to demonstrate compliance.

Sections 6.3 and 6.4 of ISO 9001:2000 draft address the requirements to provide facilities and a work environment to achieve "the conformity of the product." These particular aspects, for the most part, had previously been addressed in ISO 9001:1994 Section 9.0 (Process Control).

7.2 TRAINING

The standard requires the "assignment of trained personnel," the identification of training needs, and the provision for providing training for "all personnel performing activities affecting quality" (ISO 9001:1994 Section 4.18).

Edward Link (1997) in *An ISO 9000 Pocket Guide for Every Employee* states:

> This element of the ISO 9000 standard [1994] contains a few words but holds a high place in terms of importance to the quality system.... Training is a way to even further reduce the variation by channeling the interpretation of the documentation provided.... The proactive approach to training assures that the customer will not be a victim of training shortfalls. Business success is quite unlikely when no training takes place. (pp. 71–73)

What it takes to be "qualified" to perform specific responsibilities related to "quality" within your system must be defined. There are several factors that contribute to the actual definition of being qualified, such as education and experience. Approach this definition by thinking that every position affects quality to some degree. The standard provides the organization an opportunity to define the specific requirements. The following applies:

- Define criteria for each area of responsibility that affects quality.
- Base the criteria on the related documentation (procedures and work instructions).
- Develop orientation training for new hires.
- Provide specific training on the ISO-compliant quality management system.
- Develop qualification training for those being transferred into areas of new responsibility.
- Define requirements for identifying training needs.
- Identify required records to be maintained to demonstrate that all defined training requirements have been met.

ISO 9001:2000 expands on the training requirements by also requiring that the "competency" of associates be evaluated, documented, and addressed as appropriate for the system.

System Training

A training program should be developed to efficiently and effectively train associates in the ISO-compliant quality management system. This includes understanding the organization's quality policy and the associate's role in achieving it. All associates throughout the organization should have at least an overview of what compliance to the requirements of the ISO standard means and how this impacts the organization's overall goals. This should not only be applied during the initial implementation phase but also as an ongoing refresher program presented at defined intervals, such as once every 12 months.

Thoughts on Orientation Training

> It is vital to continue to train managers and employees in the ISO 9000 systems. As the people in the business change, an ISO 9000 indoctrination system is critical to maintaining an effective Quality System.
> —David Largey: Quality Assurance Manager, Campbell Soup Company

Defining the Training Criteria

The defined training criteria must include associates at all levels of the operation performing responsibilities that affect quality, including those in supervision and management positions. Qualifications for training must include such areas as GMP (Good Manufacturing Practices), HACCP (Hazard Analysis Critical Control Point), regulatory, hygiene, and the like. Records must be identified and maintained to demonstrate that all required training as defined in area-related procedures and work instructions are being met. An effective means to approach this is to use these documents as the primary training tool. They are an excellent vehicle for training.

This can be accomplished using a matrix designed to identify the relationship between the responsible position and the related procedures and work instructions. The documentation would be listed across the top and the job responsibilities down the side. This specific matrix would be the controlled document providing the defined requirements to be qualified for a specific position or responsibility. To provide the record, the same style matrix with the same documents listed across the top and the associate's name (title) listed in the vertical column. The date that the training was completed would be entered into the "field" that corresponds to the position/document for which the associate was trained. A sample of this matrix is illustrated in Figure 7.1.

	Job Descriptions				Work Instructions		
Name	Blender Lead (orange line) Rev 0	Past operator Rev 01	Shipping Rev. 00	Warehouse and Storage Rev 1	Ware 01 Rev 2	Nonconforming Product Rev 0	Receiving Insp Rev 0
Line workers							
John Hall	■				■	■	■
Derrick Scott	12/31/99	3/1/00		3/1/00	2/1/00		
Warehouse							
Maggie Leigh	3/23/00				9/1/00	3/1/00	3/1/00
Lilee Berry	5/1/00				9/1/00	3/1/00	3/1/00
Lucky Charm	5/1/00				9/1/00	3/1/00	3/1/00
Receiving/inspection							
Sophie Rose						3/1/00	3/1/00
Smokey Macdee							
Tester							
Janie Cox	■	■			■	■	■
Shipping							
Gary Hawks			■				
Stock Person							
Sandra Smith				■			

FIGURE 7.1 Production Associate Training Identification and Records

This type of accountability is especially useful in the processing environment where there is a constant need for cross training. The completed matrixes provide the records demonstrating that associates are qualified in their respective responsibilities. Keep in mind that associates performing relief activities must be qualified to perform these activities with records available to confirm that.

Supervision and Management Criteria

It is much more difficult to define specific training criteria for those associates that are performing less structured activities such as management and supervision. It is recommended that the specific training requirements for supervisors and management personnel to be qualified be defined in a "training plan." Based on the associate's education, experience, and previous training for the position, the training plan would be developed by the associate's immediate supervisor. Once the associate has satisfactorily completed all the tasks on the training plan, it would then be initialed and dated by the appropriate supervisor or other designee as defined in related procedures and/or work instructions. The training plan should include training in all related procedures/work instructions and be maintained as the record demonstrating compliance to all defined requirements. Supervisors and managers tend to have a variety of different backgrounds, education levels, and the like. This type of approach works well with fitting the associate's background to the qualification requirements. An example of how this approach is effective based on diverse backgrounds would be in the case of a blending operator who has been working within the department for 5 years and is now being promoted to department supervisor. That individual would have been trained in the related documents with records to prove it but may need additional training in supervisory techniques, computer software, and so forth. However, a supervisor transferring from a supervisory position with a sister company most likely has the supervisory training but would require training and exposure to the related site procedures and work instructions.

Job Description

Job descriptions are many times used as the defined training criteria. However, job descriptions by nature are very general and may not provide specific detail on exactly what is required to be qualified. It is generally best to use these as the foundation for assignment to a specific area of responsibility. Training to become qualified should include a combination of reviewing related procedures/work instructions and the successful completion of on-the-job training. Job descriptions provide information and thus must be maintained as controlled documents. The responsibility for maintaining, access, and distribution of the most current versions must be defined.

On-the-Job Training

On-the-job training can be defined as observing and performing the responsibilities of the job under supervision. Supervision is usually performed by another associate who has already been trained. At some point in performing these responsibilities, depending on how it is defined within your system, either the person doing the training or the supervisor will evaluate the associate's performance and sign off that he or she is trained. This sign off should be on a form or checklist that becomes part of the record demonstrating compliance. Records confirming that the trainer is actually qualified to perform these activities must also exist. Be careful when defining a specific time period for completing the training. Individuals learn at different paces and a defined time period may be too restrictive.

Defining Training Needs

Note that the following description should be expanded to also include the evaluation of "competency" as required by ISO 9001:2000. The term "training needs" relates to identifying ongoing training requirements. In other words, an associate trained in all the specific training for his or her area of responsibility must still be evaluated at a defined interval for any additional training needs. The requirements for identifying training needs *must* be defined. It does not mean that every associate must have a training need every time his or her training need is evaluated. Records must be maintained to confirm that training needs are evaluated and completed. Change in responsibility or some other unforeseen occurrence that may interfere with completing the activity should be noted in the record.

In some process situations, it is difficult to identify training needs. Associates are trained in their responsibilities. Needs may not arise unless the process changes, at which time the documentation will be revised and training performed on those documents. This type of situation may be defined in the quality manual with a statement such as this:

> For some specific areas of responsibilities as defined in area procedures, training needs are NOT identified. Should a defined activity change, then associates are trained through the revised area procedures and work instructions.

Many organizations address the identification of training needs through the annual employee evaluation and performance review. These types of documents generally are confidential in nature. Internal auditors should not have access to such confidential information. However, if the auditors can't access the information, then they can't confirm that all system requirements are being performed. This problem is many times avoided by having a detachable form attached to the confidential performance review that is removed and maintained as part of the training quality records. Other means may be to address training needs on an area checklist or training file. Software programs, discussed later in this section, may also be used to maintain and track this type of information.

Training Records

Records must be identified and maintained to provide the proof that all defined requirements are being met. Records must be maintained in compliance with the system's defined policy for quality records.

Identifying which documents will be the record can be challenging. These can be attendance sheets from classroom training, individual certificates, or the completed matrixes. Records can be filed by individuals, departments, or any other means useful to your process. Some systems choose to maintain all the records in the Human Resource Department; others choose to keep records with the department managers or even a combination of the two. System records may be maintained with Human Resources and records for specific area training documentation maintained in the departments. The latter provides a quick reference to supervisors during all processing hours. It also provides an effective opportunity for supervisors to document and file process training as it is being performed. The records may then transfer with the associate between departments. These preferences are left up to the system to define, but should be done in a manner that ensures associates are qualified in their respective responsible activities.

Record Retention Time

Defining the retention time for training records can be tricky because the record must be retained to confirm that training had occurred for the performance of all activities that have their own record retention time period. For example, it could be defined that training records will be kept for the "duration of employment" or even " the duration of employment plus one year." If this is the case, then the record confirming an internal auditor had been trained to perform an audit could be inadvertently destroyed prior to the internal audit record's 3-year retention time. If a trained associate completes an audit report then leaves employment the next week, as defined, his training record would be destroyed one year from the date of departure. The system for quality records defines that audit reports are maintained for 3 years. This creates a 2-year period where the system cannot demonstrate that the auditor had been trained according to defined requirements. Thus, the audit report for that 2-year period becomes void. A nonconforming situation has been created. An auditor would write that: "Records were not available to demonstrate that the internal auditor had been trained as required by defined work instructions and procedures."

Competency Training

As mentioned previously, the recent issuance of ISO 9001:2000 has gone a step further by requiring that associates are not only trained but that the record also confirms that the associate is competent in his or her area of responsibilities. This can be accomplished in many different ways. Some operations actually require that the associate take a written test. However, the most common manner that has proven both effective and useful is the "training checklist." The associate's immediate supervisor reviews and confirms competency. Completion of this is done through observing and interviewing the associate as he or she performs the required tasks. This checklist becomes part of the training record.

Grandfathering

"Grandfathering" is a term that may be used in reference to associates that have been performing their work activities prior to the implementation of the ISO system. Rather than having to go through the training exercises with associates that already know what they are doing, the system may decide to "grandfather" the associate into that position.

Be careful with this definition. The criteria for grandfathering must be clearly defined. Records must be available to confirm that grandfathered associates have met the defined criteria. Most organizations accomplish this by creating a form or checklist, listing the associate and his or her position. The supervisor signs off that the associate has been performing the responsibilities for that position in a competent manner and in compliance with all defined requirements. A time limit should be identified as part of the defined criteria. An example of how this may be defined follows:

> An associate having successfully performed specific job responsibilities 6 months prior to December 1, 2000 (the identified date for the inception of the system) will be "grandfathered" into that position. Confirmation and the record demonstrating compliance will be the completed form "TR-FF-02" by his/her immediate supervisor.

The training matrix provides an effective means of tracking the associates' training. This provides the record not only confirming that the individual associate has been grandfathered into specific responsibilities but also the related procedures and work instructions for which he or she has been trained.

Again, be careful, this doesn't mean that a 20-year associate can be grandfathered into every position he or she has done in the past 20 years. Set a realistic time frame and adhere to it. In addition, grandfathered associates will still need to be trained on the related procedures and work instructions for his or her area with evidence available (records) to confirm that this has been accomplished.

Document Revisions

Defined training requirements for procedures and work instructions most include a means to train on document revisions. Many have chosen to include in the revision history, a statement as to whether or not training for the current revision is required. Many times documents will be revised to address wording changes, typos, format, and the like, which do not change the actual activity; thus no training is required. If training is required, there must be records available to confirm that the training has been completed.

Thoughts on Training

> Properly training employees is the single most important thing a company can do to help make sure that the quality and consistency of the finished products

meet established standards. An established corporate training policy that is directed from the Human Resources Department should be in place to ensure that departmental training procedures are in line with the corporate policy. The Human Resource Department together with the individual area department managers should establish detailed training procedures that incorporate both on-the-job training and review of the current ISO system documentation pertaining to the job classification on which they are focusing. The personnel handling the departmental training should be trained on how to properly train others. The "grandfathering" process, which enabled us to get our certification initially, has become somewhat of a detriment to maintaining our certification. Many managers assumed that the "grandfathering" process limited the need for evaluating their long time employees because of their experience level. The only training they were doing was on new or revised system documentation. The need exists for periodic evaluation of all employees to assure that they are maintaining the necessary skills as directed by the quality system documentation. Much of the initial required training was designed and implemented by the department managers or supervisors. As the system matures, the absence of Human Resource directed training programs at the department level could lead to inconsistent and insufficient training of new and inexperienced employees. This, in turn, may lead to a significant amount of corrective actions whose root cause is traced to either inadequate training of employees or lack of employee discipline for not following procedures or work instructions.
—Tim Sonntag, VP Technical and Quality, Wixon Fontarome

Sylvia Garcia (Quality Assurance Manager, Domino Sugar) stated that

employee training is an area that we feel provided us with a tremendous benefit. This benefit affected both new employees and those transferred within the organization. These benefits were especially realized through

- Having the required "skills" checklist defining "qualifications" for specific job related training.
- Having written training procedures rather than relying on verbal transfer of information.
- Having the formalized training performed by fellow "trained" employees.

Training has been one of the most difficult areas to maintain since certification. There is an enormous amount of information to maintain regarding training. Documenting and keeping track of the needs of 85 people, not to mention the records, is quite daunting. This is still an area where we are trying to find a better way to do things.
—Dana Crowley, Production Manager, Danisco Sweeteners, Inc.

To most companies, ISO certification will lead to an unprecedented amount of quality system documentation that employees will need to be trained on. While increased training creates a more informed work force it can be at times a logistical nightmare for understaffed companies, companies with high employee turnover and those without the talents of good communicators. The culture change from a company who did little quality system training to one that does it often may be a difficult battle. However, once the quality system training is

completed it provides the asset of having well trained employees. Most employees see the training as a positive effort and are inclined to adopt the policies of the new quality system. Having these formal, well defined training programs allows for the delivery of a clear and consistent message to all employees time after time.
—Russ Marchiando, Quality Systems Coordinator, Wixon Fontarome

Temporary Employees

Some organizations contract with outside agencies to provide "temporary employees" to perform various duties within the system. Temporary employees that are performing responsibilities within the system that may affect quality must be qualified as defined in related procedures and work instructions with records maintained to confirm this. This may be accomplished in many ways. It is up to the system to define it. Examples would include the following:

- Provide a controlled procedure, work instruction, or pamphlet that defines all the basic requirements for training with the agency. Train a representative to perform this training and to provide the record. Specific process training would be carried out in the same manner as for permanent employees that would include training on related procedures and work instructions. Training records would be kept in the department for which the employee is working to provide easy reference.
- Train each temporary employee in exactly the same manner as the permanent employee, maintaining all records as required by the system.
- Maintain a checklist with duties defined for that particular area of responsibility. Define that the area leader will review responsibilities with the temporary employee and then document this on the checklist. This works well in systems that not only have a large turnover of temporary employees providing the quality record. It also serves as a quick reference during all hours of process operation. Many times records maintained in the Human Resources Department are not readily available to second- and third-shift supervision.

Whatever means is chosen must be clearly defined in related procedures and work instructions and supported by quality records that demonstrate compliance.

Temporary Agency as an Approved Supplier

The temporary agency must be maintained in compliance with all defined requirements for handling "approved suppliers." These agencies are providing a purchased "service" supply. The "purchasing document" must clearly define the product ordered. Many times for service supplies this document may be a "contract" that defines all the requirements. It is referenced on the purchasing document with a copy maintained in the supplier's file as the quality record. When referencing these, include the current date of the contract. Ensure that purchase documents are updated to reflect revisions. Another option may be for requirements to be supplied to the agency in a controlled procedure or work instruction that is referenced by its identification (title, number, etc.) including its revision status on the purchasing document. Periodically the responsible area manager or designee may want to confirm with the supplier that they have the referenced version. Often the auditor will perform this exercise with disappointing success rates.

Choosing a Software Program for Maintaining Training Records

Many times an organization chooses to maintain training records in a software program designed specifically for this purpose. There are many available. It is important when choosing one that prior to the final decision, an outline defining exactly what the system wants the program to provide is created. Then contact the various suppliers and request demonstrations and other appropriate literature. Be sure that the program has technical support available for the user after purchase.

Some decide that the record can be kept simply and maintained effectively with hard copy files, others use the software programs to provide varying degrees of information and records. It is your choice. As always, requirements must be clearly defined in related procedures and work instructions. Software programs used for controlling information and for maintenance of records must be controlled to ensure their content is current and accessible. This is usually defined through document control and includes documenting requirements for backing-up the software and recording that activity on an area log.

Frequently Identified Nonconformances

- A relief person is performing specific areas of responsibilities, however, records are not available to confirm that the individual has been trained to the defined requirements.
- Records are not available to confirm defined training requirements are actually being performed.
- It is not evident through the interview process that associates have been trained either in their related documents or in document revisions.
- Requirements for training on "sensory" attributes are not clearly defined nor are records available to confirm that training is being performed.
- Records confirming that associates meet defined requirements for grandfathering are not available.

- Temporary employees are not familiar with the company quality policy and his or her role in achieving it.
- The temporary agency performing the service of supplying temporary employees is not being maintained in compliance with defined requirements for approved suppliers.
- Purchasing documents (i.e., the contract or purchase order) does not clearly define the product (the service) ordered from the temporary employment agency.
- Although it is defined that training needs are evaluated, actual records confirming this are either not completed or not available during the evaluation.
- Training requirements are defined for associates within the process; however, not defined for management and supervisory associates.
- Records are not available to confirm that the software system used to maintain training records is being "backed-up" to protect its data as required in procedure DC-01 Rev 02.
- Documented procedures do not clearly define the verbally stated role of job descriptions in defining training criteria.
- Although training requirements are defined on job descriptions, these documents are not being maintained as controlled documents.

7.3 PROCESS CONTROL

Process control addresses the activities to ensure that "production, installation, and servicing" are performed under controlled conditions. When determining "controlled conditions" in the food industry the following should be applied:

- Ensure that the production is planned (scheduled) and communicated, as appropriate.
- Provide documented procedures where "the absence of which could adversely affect quality" (ISO 9001:1994, Section 4.9 a).
- Maintain the environment, buildings, plant equipment, and utilities in a suitable manner and GMP compliant.
- Ensure that personnel are adequately trained and knowledgeable (qualified) in area procedures and work instructions, GMPs, HACCP, food hygiene, pest control, and the manufacturing process.
- Maintain compliance with all reference standards and procedures.
- Monitor and control processes and process parameters.
- Ensure that a HACCP plan has been created and being followed. This involves not only compliance with all requirements for the critical control points but also the identified prerequisites programs.

- Include contingency plans for possible system failures and computer malfunctions.
- Provide a clear definition of requirements for any "special processes" that may exist within the process.
- Perform "suitable maintenance of equipment to ensure continuing process capability" (ISO 9001:1994, Section 4.9 g).
- Maintain records to demonstrate compliance to defined requirements.

Controlling the process, at first glance, may seem overwhelming. However, basically it applies to what is done within the process to ensure that product consistently meets defined specifications and the needs and expectations of the customer.

Food Guidelines (1995) clarifies this further by reminding us that controlled conditions may be demonstrated through many different means including "physical samples, photographs, written standards, and statistical sampling processes" (p. 13).

Planning the Production

Basically in the food industry, planning relates to the production schedule. The requirements for creating the schedule, revising the schedule, or whatever means the production is planned must be defined. This should include all the responsibilities, procedures, work instructions, and records. Procedures or work instructions should define who (which positions) require a copy of the schedule. This distribution should also address revised schedules and how one knows which version is the most current. The position responsible for creating the schedule or plan should maintain a copy as the quality record demonstrating compliance.

Special Processes

The standard states that "where the results of processes cannot be fully verified by subsequent inspection and testing... processes shall be carried out by qualified operators and/or shall require continuous monitoring and control of process parameters to ensure that the specified requirements are met" (ISO 9001:1994, 4.9). Special processes basically refer to those processes where the results cannot be fully verified by later inspection and testing. In other words, it cannot be determined if the product actually meets specifications until after it is used. However, once it is used it is destroyed. An air bag is a good example of this. Special processes are rare in the food industry. Some have used baking as an example; however, I have seen bakeries address this requirement by baking a sample of the product to confirm compliance before releasing the batch for packaging.

Suitable Maintenance

The organization must define the method in which suitable maintenance is performed to ensure process capability. Process capabilities may be translated to mean the ability to meet quality requirements and the needs and expectations of the customer. Many systems achieve this through a preventive maintenance program. It is recommended that this program include the distinction between critical and noncritical equipment. "Critical" equipment directly impacts quality and/or meeting the customer's needs and expectations. Technically speaking one could argue that all equipment is important, however, certain equipment could stop production. All equipment can be maintained through this program; however, it would be the critical equipment that must be addressed according to the strict definition. The system must define what it will take to meet "process capability."

Keep in mind that critical equipment will depend on the manufacturing process. A process making "product-to-inventory" with an extended shelf life period may be able to afford a down time of a few days to a week on a critical piece of equipment whereas a fluid milk or fresh bakery process with a short shelf life may not be able to afford a down time of even a few hours on a critical piece of equipment.

As stated, it is very important to distinguish between the critical and noncritical equipment. The critical should have clearly defined requirements that are substantiated by records. It is much more advantageous and useful to maintain the strict requirements on a smaller number of items then it is to assign the same criteria to every piece of equipment. This could number a thousand or more pieces of equipment in a large operation. An auditor will review the program. If the process classifies all equipment equally, than the auditor will look at it equally. Records for the ceiling fan should be just as on-time as those for the critical conveyor or pump. The auditor will have to evaluate the equipment as defined, not as he or she thinks makes sense. This is obviously an extreme example, however, it is meant to emphasize the importance of identifying equipment that must be maintained as critical to ensure process capability.

Thoughts on Preventive Maintenance Program

> The most difficult part of the ISO process was deciding which equipment we were going to include as "critical to the process" in our preventive maintenance program. Developing this list was a true challenge.
> —Naresh Modhera, Quality System Manager, Reckitt & Colman, Inc.

Having a preventive maintenance (PM) program is not specifically required; however, documentation must define how suitable maintenance is applied to ensure process capability. Although having a PM program is the most common method used, some smaller systems have chosen to achieve this by maintaining critical spare parts or critical extra tanks, fillers, and so forth. Whichever method is chosen, procedures or work instructions must clearly define the requirements with records maintained to demonstrate compliance. Records for the preventive maintenance program should include evidence to ensure that its performance is being monitored to address on-time performance.

Software Programs

Many maintenance departments maintain their program on a database. Some are purchased specifically for these functions; others are created using software programs such as Microsoft Word and Access. Whatever means is used, ensure that the database is controlled (backed-up) to protect the data. A procedure or work instruction should define the basic requirements and responsibility for using and maintaining the program.

Quality Records

Records must be available to confirm that requirements for ensuring process control are being met. This includes records that confirm that associates are qualified to perform their defined responsibilities. These records must be established and maintained in accordance with the system's quality records process.

Frequently Identified Nonconformances

- It could not be clearly demonstrated through the interview process and review of operational activities that requirements as defined in area procedures and work instructions are being followed.
- Related procedures and work instructions do not reflect the actual activities being performed.
- Although it is stated that preventive maintenance is performed, requirements are not defined in related procedures or work instructions.
- Procedures or work instructions do not define the process for monitoring and maintaining the preventive maintenance database.
- Review of the preventive maintenance records indicated that many of the scheduled activities were either not performed or performed after the actual due date with no explanation for tardiness documented.
- Records did not confirm that those responsible for performing preventive maintenance activities had been qualified (trained) as defined in related procedures and work instructions.
- Although it was stated that GMP and HACCP programs were in place, requirements necessary to meet these regulatory requirements were not defined in related system documents.

- It was stated that periodic GMP audits are performed; however, records were not available to confirm these activities as required in procedure PRC-03 Rev 2.
- It was stated that customer requirements included periodic sanitation audits with an "excellent" rating performed by AIB (American Institute of Baking). Although records confirmed that this was being done, AIB was not listed on the approved supplier list. In addition there was no evidence available to confirm results of these audits were being addressed in a timely manner.
- Requirements for performing pest control activities are not clearly defined in related documents.
- The pest control service supplier (A J Termite and Sons) was not being monitored in a manner compliant with defined purchasing requirements for approved service suppliers.
- Procedures or work instructions do not define the requirements for revising and distributing amended versions of the production schedule.
- Although it could be confirmed through the interview process that operators understood their responsibilities, related area procedures and work instructions do not clearly define the complete process parameters.

7.4 HANDLING, STORAGE, PACKAGING, PRESERVATION, AND DELIVERY

Requirements for these activities are very straightforward, addressing the protection of the product from damage, deterioration, and loss throughout the entire process. Requirements should be defined in area procedures or work instructions. This element is meant to address protection of the product throughout the process (receiving ingredients through shipping and distribution) until the responsibility is transferred to the customer. However, in the food industry there are many checks and balances in the manufacturing system that protect the product. These defined requirements by default tend to direct focus for this element on warehousing and shipping.

Edward Link (1997) in *An ISO 9000 Pocket Guide for Every Employee* supplies the following thoughts on management's role in this aspect of the system:

> The treatments provided to the storage and movement of product should be carefully considered. You cannot recover the cost of customer dissatisfaction from an insurance company. Late deliveries and the arrival of damaged goods are of no use to the customers. Assuring the requirements of the standard are met will function as the insurance against customer dissatisfaction in this category. (pp. 61–62)

The following activities must be defined:

- Methods for handling the product to protect against damage and deterioration.
- Authorization for transfer of product and shipment.
- Control of packing, packaging, and marking processes (including materials used) that ensure activities conform to specified requirements.
- Preservation and segregation of the product while under the control of the operation. (Preservation does not refer to adding preservatives to the product.)
- Protection of the quality of the product after final inspection and test depending on the customer agreement.
- Evaluation of the condition of the product in storage at defined intervals.

Approved Suppliers

Suppliers that provide shipping services (e.g., carriers) must be identified and maintained in compliance with defined requirements for the purchasing function as applied to approved suppliers. Depending on how those requirements are defined, they may be either addressed by the responsible department or information provided to purchasing. Those responsible for assigning transport or shipping companies must have ready access to the "list" of the approved shipping suppliers. They must also be trained in their related requirements for that activity. In other words, if one is responsible for evaluating these suppliers, then they must be trained in how to perform the evaluations. This may also apply to suppliers of warehousing services (i.e., storage).

The purchasing requirement for communicating via the purchasing documents the product or service required also applies. The service requirements must be clearly communicated to suppliers. In the case of suppliers for transport or shipping services, requirements may refer to shipping temperatures, trailer conditions, and the like. For suppliers of warehousing services, this could be storage temperatures, inspections, and so forth. This information must be defined in the purchasing documents, which depending on the system may be included in a contract or on a purchase order. These documents must be maintained as controlled documents and/or quality records, as appropriate. A contract that defines the requirements would be a record because it demonstrates compliance to the fact that the purchasing document defines the requirement. It would also be a controlled document because it provides information and must be of a specific version. An option to defining requirements on the purchase order or in the contract would be to do so in a controlled procedure or work instruction. This document could then be referenced by identification and revision status on the purchasing document or contract. If the supplier is required to maintain a controlled copy of the procedure or work instruction, then ensure that the process for issuing updated versions (and updating the reference on the purchasing document or contract) is clearly defined in compliance with "document control" requirements.

Storage

Designated storage areas must be used in such a manner to prevent damage and deterioration of the product while awaiting shipment. When considering methods for control of food products include such processes as warehouse storage conditions, GMP compliance, pest control, housekeeping, shelf life, storage temperatures, delivery temperatures, regulatory requirements, sanitation, and contamination risks.

Storage requirements may include temperature and humidity conditions. Depending on the importance of these requirements, these units may require calibration. Should this be the case, then ensure that units are maintained in compliance with defined requirements for the system's calibration process. Records must be available to confirm this. In addition, the system should define periodic "verification" such as recording temperature and humidity. These may be readings on a daily checklist or monitored with a recording thermometer. Frequencies and methods will depend on the importance of the requirement. In some instances, storage conditions may relate to customer requirements for ingredients or finished products.

Other requirements that may relate to handling, storage, packaging, preservation, and delivery may include

- The rotation of product.
- First-in first-out (FIFO) concept practiced.
- Inspection of trucks before loading.
- Inspection of storage areas.
- Finished product and raw ingredient storage.

Never lose sight that the goal is to protect the quality of the product and to ensure that the product remains in specification. Storage and delivery conditions must also protect the product.

Nonconforming Product

Nonconforming product may be considered any product that does not meet specifications or that for what ever reason (damage, leaking, code date, etc.) is not acceptable for use or shipment to the customer. This product should be clearly labeled as "hold" and stored in a segregated area that protects it from inadvertent use or shipment. Requirements for storage and shipping personnel as related to nonconforming product may be defined either in procedures or work instructions specific for their area of responsibility or specific for handling systemwide nonconforming products. If the latter is the method of choice, than the storage and shipping personnel must have ready access to those procedures and work instructions. It must be emphasized that nonconforming product handling activities must protect product that does not meet specifications from being used or inadvertently shipped.

Hazardous Materials

Storage requirements must also address regulatory requirements for handling hazardous materials including cleaning chemicals and any other chemicals that either might be used in the process or that may come in contact with the shipping containers. Storage of incoming materials such as glass bottles must also be defined in a manner to ensure that the final product is not jeopardized.

Delivery

Some systems define that the product may be shipped prior to final approval as long as it remains within its control. In other words, it is not actually released to the customer even though it may be in the shipping process. If this is the method of practice, then procedures and work instructions must clearly define the control the operation has on this product until its final approval. This must be substantiated by quality records, It is recommended that the operation at a defined frequency actually test the process to ensure and confirm its control. Requirements for this test, including the records that will demonstrate compliance, should be defined in appropriate area procedures and work instructions.

Edward Link (1997) in *An ISO 9000 Pocket Guide for Every Employee* wrote the following comment on delivery:

> When your company is responsible for the delivery of product (inbound or outbound), you must take the necessary steps to protect the quality of the product. Even when your company is not contractually responsible for delivery to destination, protection must be provided from final inspection and test to the point where the responsibility is transferred. (p. 61)

The point where responsibility is transferred should be defined in the quality manual.

Potential Auditor Questions

Examples of questions that the auditor may ask the shipping personnel may be:

- How do you know what is supposed to be shipped and where it is to be shipped?
- How do you know that the product is acceptable for shipping?

    The process for preparing the product, confirming its approval, and frankly getting the right product to the right customer (filling the order correctly) should be clearly defined in area procedures and work instructions.

- What must be done if a problem arises either with the product itself or that interferes with meeting the customer's requirements such as on-time delivery?

    Knowing what to do in this instance is imperative. This relates to discussions of the requirement for "amending the order" addressed in "contract review." Depending on the system, this requirement will vary. However, whatever the requirement, it must be clearly defined in area procedures and/or work instructions. The associate must be trained to understand his or her responsibility in handling this situation.

- How do you know if the customer has any special requirements? If so, how do you ensure that these are being met?"

    This relates to meeting the customer's needs and expectations. How the needs and expectations are communicated, requirements for meeting them, and what to do if an issue surfaces must all be clearly defined in the appropriate procedures or work instructions.

Frequently Identified Nonconformances

- Although procedures define requirements for storing finished product, records were not available to demonstrate compliance to these requirements.
- There is no defined method, nor is there evidence that product is being evaluated at defined intervals to ensure that it is not damaged or unacceptable.
- Suppliers of shipping services or distribution warehouses are not being identified and maintained as defined for approved suppliers.
- Storage work instruction WARE-02 Rev 02 defines that product is stored and used according to the FIFO method. However, several bags of raw ingredients with a March 2001 production code date were noted in the production area while pallets of the same product dated with a November 2000 production code were observed in the warehouse stored.
- Although it was stated verbally that trailers must be inspected for acceptability criteria prior to loading finished product, the requirements for this activity are not defined in related procedures or work instructions.
- It was noted during the evaluation that there were several pallets of cans with damaged and/or torn shrink-wrap protection. There was no means to protect this product from contamination defined within area procedures or work instructions.
- The glass container storage area is located next to bags of raw ingredients. There was evidence of several broken containers and loose glass in the area. No provision is defined for addressing these situations so as to protect against the possibility of a food safety hazard.

- It could not be demonstrated through the interview process or review of training records that associates responsible for shipping the product had been trained on the defined requirements for reporting supplier (transportation services) nonconformances.
- Approved suppliers for shipping are being maintained on an approved list in compliance with purchasing requirements; however, this information is not readily available to those personnel actually arranging the shipment of products.
- Although it was stated that that the warehouse and other storage areas were being maintained in GMP compliance, there were several violations noted during this evaluation. Also training records were not available to confirm that associates responsible for maintaining compliance status had been trained in the requirements.
- There is no evidence that associates have been trained in related purchasing activities for the use of approved suppliers of shipping and warehousing services.
- Procedures and work instructions do not clearly define the actual activities being performed by shipping personnel to inspect shipping vehicles prior to loading.
- Segregated hold areas contain product identified with both hold tags and release-for-shipment tags.
- The purchase order for warehouse services references that requirements are defined in work instruction WARE-01 Rev. 2; however, the warehousing service company could not produce this document upon request.

8
PRODUCT REALIZATION (ISO 9001:2000 SECTION 7.0)

8.1 AN OVERVIEW

Section 7.1 of ISO 9001:2000, Planning of Realization Processes, almost directly relates to ISO 9001:1994 Section 4.2.3 for quality planning. Basically, the organization will have to plan the necessary steps and actions to produce the required product. The requirements are defined in Section 7.1 a–d and do follow a logical sequence of events.

Requirements for Section 7.2 (Customer Related Processes) were for the most part addressed in ISO 9001:1994 Section 3.0 (Contract Review) and focus on ensuring that both the customer's stated and implied requirements are met. The method for communicating with the customer must be clearly defined and include communication regarding product information, order issues including amendments to the order, and handling customer feedback such as complaints.

Section 7.3 (Design and/or Development), although possibly stated in more specific terms, is very compatible to the requirements of ISO 9001:1994 Section 4.4 (Design Control). The method that this will apply to current ISO 9002:1994 approvals was discussed in Chapter 3 of this text. The current registrar should work closely with the organization on applying the "permissible exclusions." The organization's scope of approval may need to be redefined to clearly address the relationship of design and/or development to its quality management system.

Although some organizations define that the requirements for evaluation of suppliers in compliance with ISO 9001:2000 draft section 7.4 (Purchasing) have become more comprehensive than ISO 9001:1994 Section 6.0

(Purchasing), overall, it appears that the actual statement of requirements has been simplified.

Section 5.5 addresses production and service operations and actually provides the basis for some of the previous comments that the revised standard links operational activities in a manner that makes sense to the overall process. This section actually relates to a sampling of requirements previously addressed in ISO 9001:1994 Sections 4.9 (Process Control), 4.11 (Inspection, Measuring, and Test Equipment), 4.12 (Inspection and Test Status), 4.10 (Inspection and Testing), 4.15 (Handling, Storage, Packaging, Preservation, and Delivery), and in 4.19 (Servicing).

Product identification and traceability (ISO 9001:1994 Section 4.8) is now addressed in ISO 9001:2000 Draft Section 7.5.2 with no major changes in these requirements noted at this time.

Additional areas addressed in ISO 9001:2000 Draft Section 7.0 with minor changes and emphasis include customer-supplied property (referred to as customer-supplied product in ISO 9001:1994 Section 4.7), preservation of the product, validation of processes, and the control of measuring and monitoring devices (calibration).

8.2 CONTRACT REVIEW

"Contract review" may be translated to mean "order" review in the world of ISO and applies to all agreements between the customer and the organization. An ISO 9001–compliant quality management system focuses heavily on meeting the customer's needs and expectations. These requirements must be understood and it must be confirmed that they can be met before the commitment to provide the order is made. Never accept an order unless there is no doubt that that order can be completed to meet the customer's needs and expectations. This includes not only meeting defined specifications but also being able to provide the quantity on time according to customer requirements. Edward Link (1997) in *An ISO 9000 Pocket Guide for Every Employee* stated that "the strength of the company is closely linked to how well contract review is done" (p. 15).

Requirements for taking an order, whether it be via phone, fax, email, or from other means must be clearly defined in area procedures and work instructions. These must also clearly define the process for addressing and resolving any "differences" between the requested order and the organizations ability to provide it. Good examples of where differences may arise would be quantity, cost, or ship date. For example, if a customer requests 5000 cases to be delivered Friday and it is known at the time of receiving the order that only 3000 cases will be available, then this must be resolved. Resolution may be that the customer chooses to postpone the delivery until the requested quantity is available or amend the order to the available amount. This type of up-front resolution is important rather than to just accept an order and ship whatever there is at the time. The latter fails the primary focus to meet the customer's needs and expectations.

In accepting the order, be sure that customer requirements have been received, are understood, and can be met. Wherever possible, orders should specifically identify the customer's specifications for the product ordered. This may include the specification name or identification number, revision status, date, and so forth. If customer specifications are maintained, then a system must be in place to ensure the most current versions are available and being maintained as controlled external documents.

Procedures and work instructions must define the requirements for addressing amendments to the order. "Amendments" are defined as differences between the actual agreed-upon customer order and what the facility can do to fill the order. An amendment may result when an unforeseen occurrence interferes with completing the original commitment. Should something unforeseen occur, such as weather conditions interfering with shipping or equipment breakdown, the necessary activities for notifying the customer and resolving these differences must be defined. Keep in mind that different customers have different needs resulting in different types of requirements. Some want to know immediately if there are any differences, others do not require notification. These types of customer requirements must be understood and clearly defined prior to accepting the order. It may not be necessary to change anything that is being done other than to document communication between departments that play a direct role in meeting the order such as production, shipping, and storage. Ask the associates in the related departments what they would do if something happened that would interfere with completing the order. Then document the activity and identify the record that will demonstrate compliance to it.

Thoughts on Contract Review

> One of the major benefits of ISO was our ability to focus on the needs of our customers. The process of "Contract Review" was essential for us to ensure delivery of exactly what the customer wanted, and deliver it when the customer expected. The benefit to our laboratory was a dramatic increase in our credibility. We were able to give our customers the information they needed, when they needed it, with no doubts about its accuracy. This allowed them to make the timely decisions they had to, and it also increased their reliance on our services.
>
> —Victor V. Margiotta, Director of Quality, SOBE

> A major benefit that our system experienced was through meeting our customer's needs and expectations. During customer audits our individual programs/procedures/record keeping activities were impressively in-place and available upon their request. Without our system we would have accomplished this through piecemeal organization. We were able to show them that these programs are part of our quality system and that they are fully integrated into

the regular refinery routines. Because of the discipline and structure of the ISO system we performed a complete review with our customers of their product specifications. Together, we evaluated exactly what we could make and what they really needed. The overall performance to meeting customer's specifications and the reduction of out-of-specification product shipments (with customer approvals) was improved by over 85% in a four year period.

—Sylvia Garcia, Environmental and Quality Control Manager,
Domino Sugar

I am "proud" to show our procedures relative to all the typical quality audit areas to our customers. Some very favorable comments have been received as a result of this organization. The entire specification issue was always a source of major contention between Sales and Operation. Sales would promise the world, for obvious competitive reasons, but we didn't have a prayer of routinely meeting the promises, so we got close and then someone (frequently me) waved a wand and presto it was shipped. This was always a difficult position as the parameter being waved wasn't of any great significance and should not have existed in the first place. Mercifully the system forced us to get together with the customers to discuss and agree upon product requirements, our capabilities and the parameters that really mattered. We are better off as a result. The benefits of this communication and awareness are immeasurable.

—Anonymous

Quality Records

Procedures and work instructions must identify which records of "contract review" demonstrate compliance to defined requirements. These must be maintained compliant with defined requirements for "quality records."

Frequently Identified Nonconformances

- Although procedures and work instructions define that orders are taken and approved, records of these actions are not being maintained.
- Reviewed a sample of completed orders and found that they were consistently missing information.
- Through the interview process it could not be clearly demonstrated that some departments such as shipping understood the process for communicating with the customer (either directly or through customer service) should a problem arise in meeting the customer order as defined in related procedures and work instructions.
- Customer orders referenced specific specifications that were not readily available.
- Forms are completed and submitted to customer service by the sales department to communicate customer requirements; however, this activity is not clearly defined in related procedures and work instructions.

- Related area procedures and work instructions do not define the approval of customer orders and the process for addressing amendments to the order.
- Records of contract review as required by ISO 9001:1994 Section 4.3 are not being maintained.

8.3 PURCHASING

Purchasing requirements address not only purchasing but also the approval and monitoring of suppliers and activities related to on-going supplier evaluations. This has proven to be one of the major challenges in defining and implementing a compliant quality management system. Procedures must be established and maintained to ensure that product and supplies purchased meet specified requirements. Previously the ISO standard would refer to "suppliers" as "subcontractors." This referral caused confusion when initially interpreting the requirements; however, ISO 9001:2000 revised this reference. Suppliers of products and services are now refered to as "suppliers."

Compliance with the requirements for purchasing has proven to be one of the toughest implementation tasks for many organizations. Although companies usually have defined purchasing departments and activities, many times these only relate to cost and availability. These are of course very important to the business, but the ISO standard requires the documentation to address the "quality" aspects of purchasing. It is difficult for companies to stay in business if they don't maintain suppliers that provide quality products and services on a consistent basis. However many times these requirements are only informally understood and defined.

Thoughts on Purchasing

> Purchasing was by far the hardest part of implementing the system. Prior to implementation, the purchasing function was a jungle. There were poor controls and documentation requirements. Interpreting the requirements of the standard for purchasing was initially very difficult. The first procedures were very restrictive and as a result, nearly every audit showed massive nonconformances. After nearly 5 or 6 rewrites, it finally clicked (for me anyway) what the standard required. After that and very strict adherence to purchasing procedures by the purchasing coordinator, we were able to turn it around developing a process that is both compliant and useful to our company activities.
> —Dana Crowley, Production Manager, Danisco Sweeteners, Inc.

In defining these requirements, the system should differentiate between suppliers of critical and noncritical supplies. In the world of ISO compliance, "quality critical" or "quality related" generally applies to those suppliers that provide product and services that affect the quality of the product and the organization's ability to meet the customer's needs and expectations. Examples of critical supplies include raw materials (ingredients), packaging materials, processing aids, and laboratory supplies. Each department should first identify the critical supplies and services used and then proceed to define these requirements in related procedures or work instructions. Evidence must be available that the quality-critical supplies are ordered only from approved suppliers. An "approved" supplier can be defined as a supplier of critical materials and services that has successfully met all defined requirements for an approved supplier.

In many instances, a supplier may only be approved for certain products. A flavor house may actually produce a thousand different flavors but only two or three that have been approved for your operation. Be sure that procedures and work instructions clearly address this issue and that the quality records identify the specific approved products.

Emergency Situation

There may be a situation that arises where an ingredient or material must be sourced from a company that has not gone through the formal approval process. It is recommended that the related procedures and work instructions address what should be done if this occurs. This would most likely involve identifying a person of authority to make the decision along with appropriate records and traceability information that would need to be maintained to monitor the situation. Traceability is important. This provides the operation identification information just in case a problem should arise that may require product retrieval.

Service Suppliers

It is common to have responsibilities for service suppliers defined and maintained by the department manager responsible for these specific activities. Examples of this are the Operations or Quality Control Department for pest control, Laboratory Management for product testing services and laboratory calibration, and the Maintenance Department for calibration and preventive maintenance services. Other examples of critical services may be storage and distribution warehouses, water treatment companies, subcontracted manufacturers, and temporary employment agencies. Carefully assess the operation to identify all that may apply to each specific process. Service suppliers may also be managed through the purchasing department. Remember this is your system. Apply what is best overall for your operation.

Centralized or Decentralized Functions

Most operations actually perform "centralized" purchasing functions. However, depending on how the current function is being performed, requirements may be decentralized, with individual departments taking ownership for their own purchasing activities. If purchasing activities are "decentralized" and performed by representatives of many different departments, evidence (records) must be available to confirm that these responsible associates are trained in related purchasing requirements (i.e., area procedures and work instructions).

As requirements for purchasing are defined and implemented, efforts must also focus on all other related system requirements. This may include document and data control, corrective and preventive action, training and, quality records.

Supplier Approval and Monitoring

Documentation must either define the following or make reference to where this information is defined:

- Criteria for not only the approval of suppliers but also for monitoring and control through evaluation of ongoing supplier performance.
  - Focus on quality-critical suppliers.
  - Base criteria on supplier's ability to meet defined requirements.
- Criteria required to identify a quality-critical supplier.
- Type and extent of control that will be applied to the suppliers.
- Process for addressing supplier noncompliances (not meeting the defined requirements).
- Criteria/limits of noncompliances that would result in removing a supplier from approved status.
- Required objective evidence (i.e., records) that must be maintained for approved suppliers documenting their compliance.
- An "inactive" time period that would require the reevaluation of a supplier prior to use.
  - In other words, how long a period of time must pass (i.e., 1 year, 2 years, etc.) of not using the supplies or services of a supplier before it is removed from the approved supplier list, requiring that supplier to complete the approval process prior to use.

Grandfathering Suppliers into the System

When initially defining the system, many organizations choose to grandfather existing suppliers into the system. "Grandfathering" is a term that refers to suppliers that are being used at the time the system is first implemented. This is to avoid having to perform supplier approval activities on current suppliers. The criteria for acceptance of grandfathered suppliers must be clearly defined

PURCHASING   115

based on past acceptable quality performance history. In other words, suppliers used prior to system implementation may be accepted into the system on past performance whereas new suppliers would have to be evaluated against the defined approval criteria. Records must provide evidence that the grandfathered suppliers have met the defined grandfathering criteria. This can be accomplished with a checklist stating the "historical" performance status signed by the responsible buyer of the item.

Outgrowing the Grandfathered Status

Grandfathering is an activity that is applied during the implementation stage to address existing suppliers. This is the process for which they are initially approved. However, keep in mind that these same suppliers must be evaluated for performance on an ongoing basis as defined within the system. In other words, initially the supplier is approved because it was grandfathered. Once the grandfathered company passes its first evaluation, then it is approved based on the results of that evaluation, not on the fact that it was grandfathered.

Developing the Process for Approval and Monitoring Suppliers

Depending on the nature of the item referenced, the actual requirements may vary considerably. The best place to start is to develop a checklist through contributions of all those responsible for referencing and buying the item or items. Address key concerns such as ability to meet defined specifications, regulatory requirements, quality, availability of the product, and price. As appropriate, on-site supplier evaluations should be included as part of the requirement. Some organizations create an initial checklist to be completed by the potential suppliers prior to initiating the approval process. Once this information is received the process continues.

Ongoing approved supplier evaluations should include a review of the supplier's performance, any nonconformances, corrective actions, and, where appropriate, on-site audits of the supplier's facility at defined intervals. An evaluation of the adequacy and timeliness of the supplier's response to nonconformances and corrective actions should be included.

The ISO standard does not require specific actions. It is up to the operation to define what is required, adhere to these requirements, and maintain quality records that demonstrate compliance.

Approved Supplier List

The ISO standard does not specifically require that a list of approved suppliers be maintained. However, a well-designed list can be very useful in providing and tracking all required information. The list, if maintained, should include the name of the supplier, the products for which the supplier is approved, the date of original approval, the date of the most recent evaluation, and the person responsible for this supplier. Some suppliers may in fact be approved for all items or services, but some larger companies, such as a chemical supplier may only be approved for specific products. Ensure the relative information on each supplier is included in the file or on the list.

Associates responsible for ordering the items must have access to the identification of the suppliers such that, when ordering, he or she can confirm that the item is being obtained from an approved supplier for that particular product or service. Many times the current listing whether an actual list, file, software program, or other means maintained by a specific area of responsibility such as the purchasing manager. This information is then distributed as a controlled document to those who need the information to perform his or her duties. Related procedures and/or work instructions must define this distribution. An auditor may record this and follow the trail to ensure all who are required to have the document have the most current version.

Thoughts on the Approved Supplier Process

If we could do it over again, we would have searched out business examples of vendor approval/monitoring. Every plant in our company has struggled with this element and has started either by committing too much or too little versus the business need. This element is not as tied to existing practices as other elements, so it is a good candidate to standardize among plants of different backgrounds but similar needs.
—Al Gossmann, Quality Assurance Manager, Cultor Food Science

Discussing compliance to defined purchasing requirements, Glenmore Wint, purchasing clerk at Le Meridien Jamaica Pegasus, stated that "if you cannot comply you cannot supply." A short but accurate statement.

Purchasing Document

Purchasing documents may be whatever is used to order supplies such as purchase orders or contracts. These documents must include a clear description of the product ordered such as a catalog number, a specific specification number, revision, or reference to a controlled document. It must be very clear to the supplier what is being ordered. The purchasing document must be reviewed and approved to ensure that it contains all the required information (i.e., authority, specification reference, etc.). The responsibility for performing this and the required record to demonstrate compliance must also be defined in related procedures and work instructions.

Specifications Maintained for Purchasing Functions

In many instances, specifications are used to provide information for purchasing items. The specifications should be in compliance with the requirements for document control which should define the requirements for identifying and maintaining the most current version. A master list should be maintained identifying the current versions and their locations. Depending on how the process for document control is defined, this may be the responsibility of the document control coordinator or the individual departments that actually use the specifications. Remember it is your system to define.

Food Guidelines (1995) provides the following thoughts on specifications:

> Purchasing of raw materials or other primary products should also be covered by specifications. Specifications should accommodate the inherent variability of such products and encompass the need for any special controls necessary to maintain their integrity, including the requirement to meet current legislation. (p. 10)

Verification at the Supplier's Location

This applies to either a customer of the organization or the organization itself performing on-site verification at a supplier's location. These on-site activities must confirm that resulting products are being manufactured and handled as required by specifications. Requirements must be clearly defined in related documents. Performing verification on-site at the supplier's location does not absolve this supplier from the responsibility to meet all requirements.

Quality Records

Quality records must be identified and demonstrate compliance to the defined requirements. This includes records of supplier approval and supplier evaluations. Records must also confirm that purchasing documents clearly define the product ordered and are approved as required by documentation to ensure that they contain all required information. Records must be maintained in compliance with requirements for quality records.

Frequently Identified Nonconformances

- It is stated that suppliers have been grandfathered into the system, however, either the criteria for grandfathering has not been defined or it has been defined but there are no records to demonstrate that suppliers have met the criteria.
- Documentation defines that approved suppliers must be evaluated at a defined frequency; however, records are not available to confirm that this has been done as required.

- Records are incomplete with only limited evidence that suppliers are being evaluated as defined.
- Records did not always confirm that suppliers listed on various purchase orders for specific products had completed the approved supplier requirements.
- The listing for approved suppliers identifies the company but not the items for which it is approved. Allied Chemical Supply and Michigan Flavor House manufacture a wide variety of items; it was stated that those companies were only approved for some items, but documentation did not identify these specific items.
- Review of a sampling of purchasing documents did not provide complete evidence that these are being reviewed to ensure that all requirements are defined prior to ordering the product.
- Purchasing documents for those approved suppliers of services do not completely define the product (service) ordered.
- Purchasing documents for raw ingredients and packaging material do not include a complete description of the product ordered such as the specification identification and version number.
- Although the defined responsible individual for maintaining the approved supplier list has the most current version, other departments as required by related procedures and work instructions did not have the most current version.
- Specification numbers are identified on the purchase order; however do not include the version of the specification. Upon request, 3 of 5 suppliers contacted were not able to provide the specification version referenced on the related purchase order.
- Requests to the suppliers to provide a copy of the current specification or contract referenced on the purchasing documents were not successful.
- Suppliers of services such as pest control, calibration, and product testing are not identified and maintained as approved suppliers.
- Verification at the supplier's location is being conducted; however, documented procedures do not define the complete requirements for this activity.
- A specific supplier has not been used for an extended period (e.g., 3 years); however, it is still listed as an approved supplier.
- It could not be demonstrated that all departments responsible for performing purchasing activities (e.g., evaluation of suppliers) had been trained and were meeting defined requirements.
- Review of a sampling of purchase orders indicated that approximately 50% of those reviewed (10 of 20) did not clearly describe the product ordered as required by purchasing procedure PR-02 Rev 7.

8.4 CUSTOMER-SUPPLIED PRODUCT

"Customer-supplied product" is a phrase that refers to any item (e.g., ingredient, packaging, label, etc.) or service (e.g., transportation, storage, etc.) supplied by the customer in order for the organization to be able to meet the requirements of the contract. A good example of this would be if a company was co-packing fruit punch. The customer would supply the organization the frozen fruit concentrate to make the product. The organization is responsible for handling and storing the concentrate while it is under its control in a manner that meets the customer's requirements. Records must be maintained to demonstrate compliance to these requirements.

The following applies to customer-supplied product:

- "Establish and maintain documented procedures for the control of verification, storage and maintenance" (ISO 9001:1994 Section 4.7).
- Record and report to the customer "any product that is lost, damaged or is otherwise unsuitable for use" (ISO 9001:1994 Section 4.7).
- Even though the organization verifies the customer-supplied product is acceptable, it "does not absolve the customer of the responsibility to provide acceptable product" (ISO 9001:1994 Section 4.7).

Customer-supplied product may also refer to services provided by the customer. A frequent example of this may be specific transport companies or storage warehouses that the customer requires to be used for its products. Although the organization does not have control on identifying these service suppliers, the organization still has the final approval for their use. For example, the customer may request that its product be shipped via Transport Company AB. However, when the trailer is inspected it is found dirty and with a foul odor. The organization must use its judgment and reject that trailer, as appropriate. Returning to the example of the fruit concentrate, the customer may supply the drummed ingredients, but this does not mean that spoiled or substandard ingredient should be used just because the customer provided it. To reemphasize, just because it is supplied by the customer does not mean that the organization can jeopardize the finished product. Requirements for notifying and discussing unacceptable situations with the customer must be defined.

Customer-supplied product may relate to a manufacturing site that has defined its corporate office as its customer. Although the corporate office approves suppliers and instructs the location as to the source of the supplies, the organization must still define its area of control. This will most likely be through receiving (incoming) or in-process inspection. This will be the organization's opportunity to verify that material supplied by the customer meets specifications and is acceptable for use. The organization should have communication requirements for handling nonconformances with its customer (i.e., the corporate office) defined.

The description described in the previous paragraph relates somewhat to purchasing requirements. Although the organization may not have the responsibility for identifying and evaluating customer-supplied suppliers, requirements must be defined and in place for communicating performance issues for these suppliers to the customer. In some instances, the organization may communicate directly with the customer-supplied product suppliers; however, a means should still be in place to communicate performance issues to the customer.

Edward Link (1997) in *An ISO 9000 Pocket Guide for Every Employee* provides some interesting thoughts regarding customer-supplied product.

> Remembering that the customer is always right makes good business sense, but that does not mean that we close our eyes to the quality of the raw materials or supplies the customer provides and asks us to include in or use for products or services supplied by our company to them. . . . This includes raw materials, components, tools, packaging, and whatever else the customer provides for inclusion into the product or related to its manufacture and delivery. (pp. 30–31)

It Doesn't Apply?

It is quite common not to have any customer-supplied product used within a process. If this is the case, then the quality manual should make a statement to the fact that:

> At this time, this process does not use any customer supplied products or services; however, should this situation change then documented procedures will be established and maintained for verification, storage, and maintenance of these products and/or services.

There should be a means defined to evaluate the status of this situation such as annually at the management review meeting. Evidence of this evaluation should be included in the management review meeting minutes.

Frequently Identified Nonconformances

- The quality manual defines that at this time there are no customer-supplied products used in the facility; however, during the evaluation some items and services such as special packaging material or transportation services supplied by the customer were identified.
- Procedures and work instructions do not define the actual activities being performed to handle and store customer-supplied product.
- Although customer-supplied ingredients are being used, there is no defined means to report back to the customer on the performance of these suppliers. It was stated that performance issues are dealt directly with the supplier; thus the customer is not made aware of any issues.

- It could not be confirmed by record review that storage requirements for ingredients and packaging material were being adhered to as defined in the customer contract.
- Procedures and work instructions do not clearly define the actual activities being performed to evaluate the customer-supplied product to ensure that it meets specifications. Note that it was stated that this was performed during receiving inspection however, there was no evidence of this present during this evaluation.
- Requirements for using customer-supplied warehouses and shipping companies were not defined in related procedures and work instructions.

8.5 INSPECTION AND TEST STATUS

It is required that the inspection and testing status be identified throughout the process in a manner that ensures that only product that has passed all the required tests is used. Robert Peach (1997) in *ISO 9000 Handbook, Third Edition*, provides this quotation from the Victoria Group:

> This is only part of the story—the full requirement is that it should be possible to identify any element at any stage of its progress through the process, within the framework of the way that the system is established. (p. 134)

In the food industry, this is typically addressed through inspection and tests that are performed throughout the process. Typically, auditors ask questions to confirm associates know the status of the product they are using. A blending operator using raw materials may be asked how he or she knows the product has been approved for use or a filler operator may be asked how he or she knows the product is acceptable. Each process is different and how requirements are addressed will depend on its characteristics. A continuous process such as evaporation or distillation may only consist of process monitoring activities, whereas a batch process would have specific go–no-go phases. The latter most likely deals with specific test results necessary to release the product to the next phase. Typically, in the food industry, inspection and test status procedures and work instructions will reference other documents defined for such activities as process control and inspection and testing.

The following applies:

- The inspection and test status "must be identified by suitable means which [must] indicate the conformance or nonconformance of product with regard to inspection and tests performed" (ISO 9001:1994, 4.12).
- This identification and monitoring must be maintained throughout the entire process "to ensure that only product that has passed the required inspection and tests . . . is used" (ISO 9001:1994, Section 4.12).

It may be easier to understand these requirements if one thinks about applying the standard to manufacturing television sets, which may have several steps in the process. Through outward appearance, it may not be evident as to what phase of approval the product is in. In other words, if it has passed all the required tests and approved for the next process stage.

Food Guidelines (1995) summarizes these requirements with the following statement:

> There should be a clearly defined method for identification of inspection and test status to prevent inadvertent use. This may mean pass/fail, acceptable/non-acceptable, awaiting inspection notices which clearly mark particular products, or any of a number of methods suitable for other applications. (p. 16)

Edward Link (1997) in *An ISO 9000 Pocket Guide for Every Employee* describes this requirement by explaining the following:

> The simple task of identifying the inspection and test status of raw material, in process product and final product prevents inappropriate processing of raw materials or in-process product and provides further protection to avoid inadvertent shipment of rejected product to a customer. Cost and customer satisfaction levels should not fall victim to something so easily prevented. (p. 48)

Some systems use a material handling software program that requires approval for product release to be physically entered into the database. The database actually tracks the status of the product through various stages of the process. As with all software programs, ensure that the database is controlled (backed-up).

Frequently Identified Nonconformances
- Product, staged for production, had completed various approval inspections; however, it was not evident nor did operators know which product had passed the required tests and was ready for transfer to the next stage.
- It was defined that product must pass certain quality checks before being released to the filling machines. Filling operators stated that they would review laboratory tests before physically releasing the product to the filler; however, records did not provide evidence that this was being performed. (Note that verbal or visual approvals should be documented on a record to confirm that that activity had been performed.)

8.6 PRODUCT IDENTIFICATION AND TRACEABILITY

Requirements for product identification and traceability are basically straightforward. Where necessary, ingredients and finished products must be identified from receipt through all stages of production and delivery. This should be done in such a manner that product can be retrieved should a problem occur. Responsibilities for performing and monitoring the requirements must also be defined.

Robert Peach (1997) in the *ISO 9000 Handbook, Third Edition*, cautions not to confuse traceability and identification. Identification is defined as "any suitable means that allows an organization to trace a service or [product] whereas traceability is the ability to trace the history and delivery of a service or [product] at any point while the organization has responsibility" (p. 428).

Peach further quotes The Victoria Group stating that

> the level of traceability required is left to the discretion of the company unless specifically called for under contractual obligations, regulatory requirements or industry norms. . . . The company must decide, document its decision, and then adhere to it. Companies should realize that if there is an industry norm for identification and traceability, auditors would expect to see that norm being followed. If there is a regulatory requirement, companies are required to follow it (p. 107)

These activities are particularly significant for the food industry. Typically food manufacturers have been identifying their products from raw materials through the finished product for many years. This is important should an issue occur that requires a product recall from the marketplace.

The process usually begins with recording the lot number and supplier of the raw ingredient, then assigning an internal lot or batch number to the in-process product. Finished products are usually identified and tracked via the date code. Records for the production of a specific finished product date code should trace/link to the lot or batch numbers of the ingredients used in the product.

Requirements for traceability relate to what most in the food industry refer to as the "recall" process. This is, of course, very important in food manufacturing operations. Many operations have an existing program that they incorporate into procedures and work instructions. These documents define the responsibilities along with the quality records required to demonstrate compliance to defined requirements.

It is recommended that "mock" recalls be performed at defined frequencies to monitor the effectiveness of the process. A mock recall is a practice recall exercise. A request is made for information on the locations of the product with a specific code date. The date chosen should be such that the majority of the product has been distributed. It would not be a very effective exercise if the product had been recently produced and still in the warehouse. Records should include the total number produced and the total number accounted for through the distribution chain. This exercise should also be applied to raw ingredients. Pick a lot code of an ingredient and then track all the products that used the ingredient.

The frequency of the mock recalls should be dependent on what is felt is adequate to monitor and test the process. However, it is recommended that the mock recalls be performed at a minimum of once every 6 months. It may be necessary to perform these more frequently during process implementation and until consistent positive results are attained.

One hopes that the operation is never put in a position where this process has to be used for a real recall situation. The practice recall exercises are an invaluable tool to test the system. They provide a good foundation and learning curve. Results of these exercises should be maintained as quality records demonstrating compliance and proof of the status of the process. Any problems identified should be documented and addressed through the corrective–preventive action process. It is very likely that during the certification assessment the auditor will request that a mock recall be performed. He or she will choose the lot numbers at random and make the request.

Of course, depending on the product being manufactured, identification and traceability activities will vary. The organization must define the method it feels is the best for its process. It is up to the processor to know the operation, know what is required, and then define these requirements appropriately in procedures and work instructions.

The system does not demand the impossible, but it does require that, as appropriate, requirements be defined. An example of one situation may be in receiving bulk materials. Raw milk is tested, approved, and received into a bulk silo. The loads going into the silo are identified, but once blended, unique identification is lost. The system in this case would define a method that would provide an estimate as to what loads were received and when the silo was actually used. In some processes, retrieving may involve a lot of "extra" product, but by design, it must be done that way.

The ISO standard is quite clear that all requirements of the customer must be met. Thus, if a specific customer requires a specific process or identification method, then the organization must adhere to the requirement as per the agreed contract. Regulatory and industry standard requirements must be identified with requirements defined and records maintained to demonstrate compliance.

Quality Records

To further emphasize, records must be maintained to confirm that the identification and traceability documentation demonstrates compliance to all defined requirements. Complete and accurate records are very important and will in an emergency recall situation make the difference between success and failure. A recall is a serious matter; however, it becomes even more serious if all the product of concern cannot be accounted for and located in a timely manner. This may create not only the potential for a disastrous food safety issue, but also it could have a direct impact on the business of the organization.

Edward Link (1997) in *An ISO 9000 Pocket Guide for Every Employee* provides a very good description regarding employee responsibilities and quality records:

> All employees responsible for creating records that include information on traceability should do so with the utmost care. Our hope is that these records never have to be used. In the event that they are required, a lot could depend on their accuracy. (p. 35)

Frequently Identified Nonconformances

- Documentation did not define the requirements for identifying and tracing product from raw ingredients through processing and distribution.
- Although it was defined that the system would be tested every 6 months to ensure its effectiveness, records were not available to confirm that this activity had been performed at the required interval.
- It could not be clearly demonstrated through review of training records that responsible associates, as identified in related procedures and work instructions, had been qualified to perform the defined requirements.
- Review of a sampling of records for batch preparation did not provide evidence that lot numbers were being recorded as required by area procedures and work instructions.
- A mock or practice recall was performed using finished product code dated March 1, 2000. Only a limited percentage of the total product produced could be accounted for and located.
- Documentation requires that the manufacturer's lot numbers be recorded for all ingredients upon receipt; however, records were not available to confirm this.

8.7 CONTROL OF NONCONFORMING PRODUCT

"Nonconforming" product is defined as product that does not meet specifications. The term may be used in relation to such items as raw ingredients, packaging, in-process product, finished product, and customer complaints. In the food industry, control of nonconforming product has been addressed for many years through "hold" procedures. Procedures and work instructions must define the requirements for handling nonconforming product in such a manner to protect against its inadvertent or unintended use. Every effort should be made to identify and segregate the nonconforming situation as soon after it occurs as possible. Commonly, manufacturers will have a designated hold area or use a special tape to separate nonconforming products from acceptable ones. This tape is very similar to police crime tape but labeled "hold" or "do not use." In addition, ensure that, as necessary, the procedures and work instructions link/make reference to recall procedures just in case nonconforming product is released before being identified.

The following must be defined:

- The "identification, documentation [records], evaluations, segregation (when practical), disposition, ... and [process] for notification to the functions concerned" (ISO 9001:1994 Section 4.13.1).
- The "responsibility for review and authority for the disposition of nonconforming product shall be defined" (ISO 9001:1994, Section 4.13.2).

What is to be done with the product (the disposition) may include such actions as reworking, regrading, destroying, or donating. Some systems may have a waiver release process, which allows a responsible individual or designee after evaluating the nonconformance to issue a waiver for release of the product. The justification for this action must be documented as part of the permanent record.

In its continual focus to meet the customer's requirements, the standard states that if required by a customer contract, results of a nonconforming situation will "be reported for concession to the customer or customer's representative" (ISO 9001:1994, 4.13.2).

Most systems document the nonconformance on individual forms that provide all the pertinent information on the situation, including its final disposition. In addition, they also frequently use a "nonconforming log" to track each occurrence. This log provides an excellent process monitoring tool. Should an instance or a group of instances (trend) result in a formal corrective–preventive action, then the record of nonconformance should identify the appropriate identification (number, etc.) of the corrective–preventive action. Records must be maintained in compliance with the system's quality record process to demonstrate that all activities are being performed to meet defined requirements.

Thoughts on Nonconforming Product

> This section states that you must have means to identify, segregate, review, and disposition product that does not conform to requirements. Although we had systems in place before ISO, this is a very critical area and ISO just enhances the need to do it correctly.
>
> —Bill Lockwood, Package Quality Manager, Hiram Walker & Sons, Ltd.

Edward Link (1997) in *An ISO 9000 Pocket Guide for Every Employee* describes management's role in relation to nonconforming product:

> It almost goes without saying that management must first advocate the use of processes, techniques, etc. that contribute to the control and improvement of the

process in order to minimize the size of your company's nonconforming material problem. Recognizing that there is and will likely always be some nonconforming material to deal with, management needs to be confident that the nonconforming material procedures prevent escape of nonconforming material and are linked very strongly to corrective and preventive action methods. Unexplainable escalating rates of customer complaints and nonconforming material provide hints to auditors that problems might not be receiving the appropriate treatments. (pp. 51–52)

Frequently Identified Nonconformances

- Records do not provide evidence of the disposition for product identified as nonconforming on in-process and finished product testing logs.
- Pallets of product were noted in several areas of the warehouse with both a "hold" sticker and an "available for shipment" label.
- Records indicate that product has tested out of specification; however, there is no evidence that this product was placed on hold, evaluated, and dispositioned as required by area procedures and work instructions.
- Review of the nonconforming log indicated several trends in product nonconformances; however, there is no evidence that these occurrences are being reviewed, trended, and, as appropriate, addressed through the formal corrective and preventive action process.

8.8 DESIGN CONTROL (DESIGN AND DEVELOPMENT)

Design development is a logical approach to creating new items. An analogy to building a new house relates well to the overall concept. First, one must know exactly what one wants, decide where one is going to build it, draw up the architectural plans, and coordinate teams of subcontractors (plumbing, masonry, roofing, etc.). House building is a step-by-step process requiring numerous interactions between many groups with diverse backgrounds. A cement slab must be correctly poured before the walls are erected. Precision building and timely teamwork will effect the entire product. It would be useless for the roofer to show up ready to add the roof the first day of groundbreaking. Many of these steps relate to other steps in the process and must be coordinated in a simple, but effective manner.

Typically in the food industry, design control relates to those departments that actually create something new, such as engineering or research and development groups. Let's focus on the research and development department of a juice company. Marketing may request a new product, but in doing so it must specifically describe the product it wants developed. A request for a new grape juice product would be incomplete without some specific information. Information such as what percentage of juice, what price range, processing treatment (hot packed or chilled), and shelf life is necessary to even begin the project. One would not merely tell a builder "build me a house" without providing specific plans.

The design or development of a new product or service requires a distinct plan be created. Requirements must address the control and verification of product design to ensure that it meets defined requirements. The requirements for the complete function must be clearly defined in procedures and work instructions. These must include the identification of responsibilities and the objective evidence (records) to be maintained to demonstrate compliance. As the following information is reviewed, think about its application to our personal projects such as building a house.

Develop the Design Plan

Identify what is to be designed; then develop a plan to accomplish this. The plan should include identifying the required work activities, verification requirements, and the assignment of responsibilities. "The design and development activities shall be assigned to qualified personnel equipped with adequate resources. The plans shall be updated as the design evolves" (ISO 9001:1994, 4.4.2).

Research and development departments usually assign individual projects to a specific food technologist for development. The system may define that it is the responsibility of that developer to document the plan and subsequent compliance activities.

Identify the Organizational and Technical Interfaces

An "interface" may be defined as the interaction between different departments or groups that have the final outcome of the project in common. All the related interfaces must be defined. In other words, the individual departments and groups involved with the project and at which stages their interactions are required must be identified. Examples of this for a food manufacturing project may include the following:

- Engineering to ensure that the required equipment is either available or can be accessed.
- Quality assurance to ensure that the product requirements are defined and can be met.
- Production to ensure that requirements are realistic and can be manufactured.
- Distribution to ensure all shipping and storage requirements can be met.

Although it may have to be revised as the design process proceeds, it is important to not only initially identify the various stages for which specific interfaces may be required but to also gain commitment from the various interfacing groups for their realistic and timely involvement. The design of a juice product would likely require several interface meetings not only to ensure that the product being created is in fact the one requested but also that the product's design is realistic and attainable. When setting up this type of program, it is necessary to clearly communicate the required interface activities in a manner that is useful and effective.

The Design Input

In other words, what are the specific requirements for the product? Robert Peach (1997) in *ISO 9000 Handbook, Third Edition*, provides an excellent summary of this aspect of design control.

> Design inputs are usually in the form of product performance specifications or product descriptions with specifications. The requirements include the following:
>
> - Identify all design input requirements pertinent to the product
> - Review the selection for adequacy
> - Resolve incomplete, ambiguous or conflicting requirements. (p. 80)

The information defined for design input should be prepared in a manner that provides the developer a useful guide that can be applied throughout the design control process. It should include any requirements outlined in customer contracts, industry standards, composition, shelf life, packaging, price, regulatory, and other legal requirements related to the specific product and/or industry.

The Design Output

The design outputs must be defined in a manner that ensures that the product

- Meets the design requirements (design inputs).
- Is able to be verified and validated.
- Identifies characteristics crucial for safe and proper product function such as manufacturing, storage, handling, and distribution.

Andrew Bolton (1999) in *Quality Management Systems for the Food Industry* states:

> In a food manufacturing operation the outputs may well be raw and packaging material specifications; product specifications and key quality attributes; artwork design; [and] process specifications, including process quality control standards. (p. 51)

The Design Review

The review of design results must be planned and include representatives from all functions involved in the design. Formal milestone-type meetings should be conducted for these reviews. These meetings must be documented with the records maintained in compliance with the quality record requirements.

Design Verification and Validation

The requirements for the verification and validation process should be identified. This is important to know from the beginning since it provides the developer an essential part of the structured focus.

"Verification" confirms that the product meets the designed criteria. In other words, the design output meets design input requirements. The product developed is the product requested. Verification should be performed at various stages of the design development process as defined in the design plan. It may include such activities as challenge testing, shelf life determination, customer acceptance, or some other method defined in the design plan.

"Validation" confirms that the product developed is the correct product for the application. The product meets defined user needs. Some typical examples of design validation would include test marketing, evaluating a trial production run, or consumer test panels.

Returning to our example for developing the fruit punch product, marketing might request a fruit punch, burgundy colored. Verification confirms that the fruit punch is actually burgundy in color. Validation evaluates and confirms that the burgundy color was the right color for consumer acceptance.

Design Changes

The requirements for documenting design changes and revisions must be defined. These should be in a manner that best promotes the overall effectiveness of the entire design process. Design changes should consider the potential risk to product quality that would occur as a result of the change. "All design changes and modifications shall be identified, documented, reviewed, and approved by authorized personnel before their implementation" (ISO 9001:1994, Section 4.4.9).

Quality Records

The records to be maintained as quality records must be identified. These should include the design plan, verification activities, design review meeting minutes, and any other appropriate proof that will demonstrate compliance to specified requirements.

DESIGN CONTROL (DESIGN AND DEVELOPMENT)

The Purpose

ISO 9004–1:1994 Clause 8 provides an excellent focus on the purpose of design control activities. It states that the resulting design function should

> result in a product that provides customer satisfaction at an acceptable price that gives a satisfactory financial return for the proposed production, installation, commissioning, or operational conditions.

Thoughts on Design Control

> Design Control is a one of the primary benefits from ISO certification. It assures the output of a process is driven by the needs of the customer. It is an essential component in the creation of a "service based" product development process. It promotes a logical sequence of events: inputs > process > outputs.
> —Jim Murphy, Manager of Design Process and Validation, The Dannon Company

> Although not required in ISO 9002:1994, it is a good practice to plan your new products to help them be successful. We have this documented in our quality planning section.
> —Bill Lockwood, Package Quality Manager, Hiram Walker & Sons, Ltd.

The situation described below should not be an issue with the revised ISO 9001:2000 since whether an organization can exclude its research and development activities will be evaluated closely through the "permissible exclusion" activities:

> A food business that has Research and Development on-site should develop ISO 9001 from the beginning rather than integrating the design element into an existing ISO 9002 system after initial accreditation. Most of our manufacturing facilities did not include R&D in the initial development program for ISO and we lost opportunities that would have been defined by a more disciplined development process.
> —David Largey, Quality Assurance Manager, Campbell Soup Company

Design control requirements basically focus on commonsense-type activities that logically should be followed to create a successful new product. There are so many new food products continually introduced, one would think following this format is an everyday occurrence. Research and development departments may be performing all the activities, just informally. This structured discipline, when formally defined and adhered to, promotes consistency between developers. In some instances, a product may not include all the steps, but this would be documented in the project plan, as appropriate. The requirements should be structured to allow flexibility such that they can be applied consistently to various types of development projects.

Edward Link (1997) in *An ISO 9000 Pocket Guide for Every Employee* provides the following excellent description of the role of management:

> The management of your company needs to encourage the use of the structure provided by an ISO 9000 compliant Design and Development procedure. The planning required, the cross-functional aspects, the periodic review and the required testing all contribute to reducing the time required to deliver successful designs. Designs that don't fulfill customer requirements or are unavailable at the right time affect business. (p. 18)

Frequently Identified Nonconformances

- Records are not available to demonstrate that the design process is being documented as required by procedures and work instructions.
- Evidence is not available to confirm that all interface activities have been performed as identified in the plan.

8.9 SERVICING

Servicing refers to after-sales or warranty-type activities. When required by contract, procedures defining the servicing requirements must be established and maintained. There must be evidence that servicing activities meet specification. Servicing-type activities may be rare in food-related processes; however, some examples may include maintenance of refrigeration equipment, vending machines, and juice dispensers used in food service.

It is common in the food industry for a system that does not perform servicing-type activities to simply state in its quality manual that "at this time servicing-type activities are not being performed and are not required within the operation." Do not state that this does not apply because the ISO standard does address it; however, the system at this time does not perform these activities. A means should be defined to periodically review the status of the servicing activity to ensure that it has not changed. This could be done annually at a management review meeting. Include this as part of the management review agenda with results of the discussion recorded in the minutes.

Frequently Identified Nonconformances

- The quality manual states that servicing-type activities are not being performed; however, there is no defined means to periodically monitor the status of this.
- The quality manual states that servicing-type activities are not performed and that this status will be evaluated at least annually during the management review meeting; however, there was no objective evidence

(review of minutes) available to demonstrate that this had been performed.
- Records were not available to confirm that servicing activities as required by customer contract (QU-04/September 01, 2000) were being performed.
- Review of a customer contract states that the organization will provide and maintain refrigerated storage units for product display; however, related documentation (quality manual) states that these activities are not being performed.

9
MEASUREMENT AND ANALYSIS (ISO 9001:2000 SECTION 8.0)

9.1 AN OVERVIEW

Overall Section 8.0 (Measurement, Analysis, and Improvement) varies the most from defined requirements in ISO 9001:1994. Its primary focus is on continual improvement of the quality management system.

The requirements for planning state that "the organization shall define, plan and implement the measurement and monitoring activities needed to assure conformity and achieve improvement . . ." (ISO 9001:2000 draft, Section 8.1). Basically, this requires that the organization identify what must be "monitored" and "measured" and then define the means to accomplish it. In other words, "planning" can be defined as those activities required to implement measurement and monitoring activities to not only ensure conformity to requirements but to promote improvement throughout the quality management system.

ISO 9001:2000 draft Section 8.2.1 (Customer Satisfaction) is a new requirement that deals with the monitoring of customer satisfaction and dissatisfaction as one of the "measurable" objectives of the quality management system. This is meant to determine "how" the customer actually feels about the organization's ability to meet the his or her needs and expectations. The organization must define how these requirements will be met. In evaluating customer satisfaction compared to dissatisfaction, a process must be defined to measure both what satisfies the customer and what does not.

Requirements for the internal audit activities are defined in ISO 9001:2000 draft Section 8.2.2. Although the wording, in some instances, may be slightly different, the requirements for this activity did not really appear to change significantly.

ISO 9001:2000 draft Sections 8.2.3 and 8.2.4 (Measurement and Monitoring of Processes and Measurement and Monitoring of Products, respectively) basically require that the means to ensure that requirements for both processes and products have been identified and are being achieved.

ISO 9001:2000 draft Section 8.3 addresses the requirements for handling nonconforming product. Other than some wording changes, the basics of this relates to ISO 9001:1994 Section 4.13. Overall this requirement may be considered less rigid than that defined in ISO 9001:1994.

ISO 9001:2000 draft Section 8.4, titled Analysis of Data, requires that appropriate data be collected and analyzed to evaluate the suitability and effectiveness of the quality management system and to provide ways to identify opportunities of improvement.

ISO 9001:2000 draft Section 8.5.1, Improvement, focuses on identifying opportunities for top management to be proactive. The purpose of this section is to provide evidence that top management is evaluating the overall quality management system for improvement opportunities.

ISO 9001:2000 draft Section 8.5.2 addresses corrective action and focuses on identifying and addressing issues of nonconformances to correct the situation and prevent it from happening again. ISO 9001:2000 draft Section 8.5.3 defines the requirements for preventive actions. Preventive actions must identify and address potential nonconformances to detect, analyse, and prevent their occurrence.

9.2 INSPECTION AND TESTING

ISO 9001 addresses the requirements for performing receiving, in-process, and finished product inspection and testing. These requirements are very straightforward and meant to ensure that product requirements are met. Management and associates must strive to define the amount of inspection and testing activities that will in a practical and useful manner provide the assurance that product is in compliance. Excess inspection adds cost to the operation, but insufficient testing will result in nonconforming product being released to the customer. Keep in mind that inspection and testing is meant to demonstrate compliance to defined specifications. It is not meant to control the process or to inspect the quality. Activities performed in compliance with process control should control the process to ensure that product manufactured as defined in the quality plans is indeed in conformance.

Receiving Inspection

Procedures and work instructions should be available to ensure that raw ingredients and materials are verified to confirm that they meet defined requirements prior to their use. Verification can be defined by suppliers by whatever means fits the operation. This could be through actual testing, visual approval, or verification via certificates of compliance or certificates of analysis. A certificate of analysis (COA) provides the analysis of test results. The certificate of compliance (COC) provides the test results along with a statement confirming that the product conforms to all the specified requirements. Whichever is the implemented, complete requirements must be defined. Evidence (records) must be available to demonstrate compliance.

If the supplier is required to provide either a COC or COA with each shipment or for each lot, it must be verified that this is indeed happening and that these documents clearly identify all the required information. If a specification is referenced by the supplier, it must be ensured that the proper version of that specification is being maintained by the supplier. Procedures and/or work instructions should define what to do [i.e., reject the load] should the COC or COA not be provided.

Robert Peach (1997) in *ISO 9000 Handbook, Third Edition*, provides an excellent description of the requirements from ISO 9004-1:1994 for performing the receiving inspection requirements and maintaining related records:

> Clause 9.7, Receiving Inspection Planning and Control, notes that the "extent to which receiving inspection will be performed should be carefully planned ... The level of inspection should be selected so as to balance the costs of inspection against the consequences of inadequate inspection." Clause 9.8, Receiving Quality Records, stresses that appropriate records should be kept to "ensure the availability of historical data to assess [supplier] performance and quality trends." Companies should also consider maintaining "records of lot identification for purposes of traceability." (p. 119)

Maintaining the records of performance and the identification for lot traceability are excellent examples of how various ISO requirements relate to each other. The standard requires that incoming material required for "urgent production needs" prior to its approval be clearly identified and tracked in a manner that it can be retrieved if necessary. Incoming ingredients should not be used without approval unless in a very extreme situation. Procedures and work instructions should clearly define the responsibilities for making these decisions and the manner in which the items will be identified and tracked to permit recall.

In-process Testing

In-process testing deals with testing the product or service while it is being manufactured or performed. These are the tests and inspection activities that confirm acceptance of the item during its process. Requirements for these activities should be defined in related procedures and work instructions. Activities should confirm that all required inspections and tests are performed and that the product meets requirements before it is released to the next step in the process. An exception may be releasing under "positive recall," as described previously for receiving inspection.

Final Inspection and Testing

All required tests and inspections must be performed on the product or service prior to its release. In other words, finished product must comply with all defined requirements prior to its release from the control of the organization. Some organizations may ship the product to distribution centers or customer warehouses while waiting for results from microtesting. In these situations, procedures and work instructions must clearly define the organization's control for "hold" and "retrieval" should the product not pass these tests. The process for releasing the product once acceptance criteria is confirmed must also be defined.

Inspection and Test Records

The standard does require that records be maintained to confirm that all required tests are performed and product has either passed or failed these checks. Product that has not met its requirements should be addressed as defined in the requirements for nonconforming product. Records should be maintained in compliance with defined requirements for quality records. Ensure that all records required by regulatory agencies or the customer are being maintained in an appropriate manner (readily accessible, etc.).

Addressing the Requirements

Required testing and inspection activities should be defined by the operation. The ISO standard does not specifically state what to do. The approach to meeting these requirements will depend on the specific process. Some processes are continuous such that in-process tests, if performed, are done as a process monitoring activity. Adjustments may be made to the process as a result of the test results; however, these are more for fine tuning the process than to verify that it meets defined specifications. Other types of processes depend directly on the approval at each stage. For example, a batch of product is not released for filling and packaging until all required tests are performed and it is confirmed that the product meets specifications. Related procedures and work instructions (the quality plans) should clearly identify and define required activities. These must be performed as defined with records maintained to demonstrate compliance.

Never underestimate the importance in proving compliance to defined specifications. However, ensure that these requirements are clearly defined in a manner that is useful for the process. Do not define a specification that is unattainable and then operate out-of-specification indefinitely. Many organizations "shoot themselves in the foot" creating nonconforming situations. By defining what they would like the specification to be, knowing that for one reason or another (process capability, design, etc.) that they will never achieve it.

Frequently Identified Nonconformances

- Review of records for inspection activities did not demonstrate that all required tests as defined in work instruction LA-02 Rev 2 were being performed.
- Procedures require that a certificate of analysis is received for all incoming products, however, review of records indicated that in several instances items were received without these documents.
- Review of results from specific tests indicated that they were frequently not meeting the defined requirements; however, there was no indication that any actions were performed to bring the product into specification prior to use.
- Procedures and work instructions did not clearly define the requirements for confirming that product had passed all required tests prior to release for shipping.
- Although it was stated that approval was received verbally, there was no defined means nor were there records to confirm that product had passed all in-process tests prior to release for filling and packaging.
- There was no evidence nor was it defined in related procedures and work instructions that those responsible for receiving raw ingredients and materials were trained and understood requirements for nonconforming product identified and handled within their control.
- Ingredients stacked on the loading dock awaiting approval were being stored with those products already approved without any clear identification or distinction to protect against the inadvertent use of the unapproved product.
- Although records were available confirming that nonconforming raw materials had been rejected, there was no means defined to communicate this information to those responsible for evaluating supplier performance.
- Review of a sample of COAs indicated that several of these did not provide complete evidence that all required tests were being performed.
- Procedures and work instructions did not define the requirements for receiving packaging materials.
- Records did not provide evidence that lot numbers were being recorded for all raw materials received as required by procedure REC-01 Rev 02.

9.3 CONTROL OF INSPECTION, MEASURING, AND TEST EQUIPMENT

The phrase "controlling inspection, measuring and test equipment" can be more clearly translated into the term "calibration." The calibration of equipment that is used to demonstrate "the conformance of product to specified requirements" (ISO 9001:1994 Section 4.11.1) will be referred to as "critical equipment" from this point further. It is essential to the overall process that this equipment perform to a degree of accuracy necessary to ensure that the product is within specification and meeting the customer's requirements.

When considering this aspect of the standard, the first of many critical requirements is to identify all equipment that is used to "demonstrate conformance to defined requirements." Keep in mind, this does not require *all* equipment used to measure and test. There is a difference!

Each process is different, and it is up to the management of the system to identify specifically the equipment that is used to demonstrate "the conformance of product to specified requirements." It is very important to understand that this does not mean that some equipment doesn't have to be accurate. It merely separates a specific category that will require specific actions to maintain compliance. Some use the same system for maintaining the "noncritical" equipment with just less stringent actions, tolerances, and the like defined.

A general rule of thumb is that if the equipment is used to test or monitor an activity that is later confirmed by another test, then only the last test that confirms compliance has to be performed on a "calibrated" piece of equipment. A good example of this would be a hand-held refractometer used in blending fruit punch. These refractometers are used by the blending operator to monitor the blend, once it is completed, the official reading is measured in the laboratory on a calibrated refractometer. Do not misunderstand this information to mean that the hand-held refractometers do not need to be accurate. It is important to the efficiency of the operation that these are as accurate as possible. They can be checked and adjusted routinely as defined in the area procedures or work instructions, but in doing so they do not require the degree of scrutiny, traceability, and documentation that are required for records of critical equipment.

Thoughts on Calibration

We never had a regular calibration schedule for most of our measuring and testing equipment. Often the calibration that was done was "hit or miss" on an infrequent basis. I can just imagine the amount of product that was given away for free before this system was put into place. We had a great deal of resistance from lab managers when they were asked to establish regular calibration programs for their equipment. Many could not understand the need to calibrate equipment to the extent of traceability to a national standard. Our regular calibration schedule is often accompanied by regular cleaning and preventive maintenance for the equipment ensuring its proper functioning.
—Russ Marchiando, Quality Systems Coordinator, Wixon Fontarome

Critical Versus Noncritical

Each system should identify its critical and noncritical equipment. In doing so, there should be a conscientious thought pattern and evaluation applied. Robert Peach in *ISO 9000 Handbook, Third Edition*, lists the following as criteria to use as guidance when making these decisions:

> What the measurement will be used for, the required tolerance versus equipment capability, the ruggedness of the equipment, working conditions, frequency of use, possible malfunctions, whether the measurements will be supported by other data, [and] whether the measurements will be used to support a specification or claim. (p. 129)

Also are there any specific measurements required by the customer?

The process for classifying the requirement should be documented. Although not required, the noncritical equipment should also be identified in a manner that distinguishes it from equipment classified as critical. This alleviates confusion. Identification can be done effectively by applying a different color tag or sticker. Associates must be trained to understand the difference. Procedures or work instructions must define the differences.

Calibration of the critical testing equipment can be maintained through the preventive maintenance program. Don't mistake this comment to mean that preventive maintenance is required for the calibrated equipment. Generally, the preventive maintenance program is designed to notify when a task is due. This is an ideal means to trigger requests for calibration to ensure that the calibration time frame does not expire. This type of program provides a well-structured system for maintaining the schedules and records.

Thoughts on Critical Versus Noncritical Equipment

> Calibration of equipment that is used to check product conformity to specifications can often be overlooked in non-ISO quality systems. This was somewhat true in our case before we implemented our ISO 9001 system. While we had programs in place for calibrating some of our measuring and test equipment like balances and scales, pH meters and other analytical equipment, we had overlooked the calibration of some production and lab equipment such a thermometers, humidity monitoring equipment and product screening equipment. When we implemented our ISO quality system, the requirements of the standard not only helped us define what equipment should be calibrated but it also provided a structured approach to assuring applicable equipment is in calibration by defining a schedule of when specific equipment is to be calibrated. The ISO standards also dictated the necessity of calibrating inspection, measuring, and test equipment in a way that is traceable to a national standard (where applicable). This was not always the case with our pre-ISO calibration program. In summary, implementation of an ISO 9001 quality system has enabled us to implement a well structured equipment calibration system that gives us much better confidence that our products are conforming to product specifications then we

have had in the past. In addition, our current calibration program is much better documented and the calibration status of our "critical" equipment is much more visible than before our ISO implementation effort.
—Tim Sonntag, VP Technical and Quality, Wixon Fontarome

To reiterate, many times the system will include all equipment and not designate the "critical," making the system difficult to maintain. It is very important to apply the stringent requirements to that equipment that effects the quality of the product or demonstrates compliance to defined requirements. As stated previously the equipment that is used for process monitoring is still very important and the means for maintaining that should also be defined appropriately with records available to demonstrate compliance. Requirements for process monitoring equipment may be maintained in the preventive maintenance program.

The Requirements

Procedures and work instructions should address the following requirements for critical equipment:

- Method for unique identification (equipment numbers, serial numbers, etc.).
- Method for which the calibration status will be identified such that it is readily evident to the associates using the equipment (labels, plaques, logs, etc.).
- Frequency of tests required to ensure accuracy and suitability.
- Methods for testing (reference appropriate work instructions).
- Identification of the standard (i.e., national, international, or known) to be used.
- Required training for associates performing the calibration activities.
- Tolerances related to the required degree of accuracy for the tests being performed. (In other words, if a scale is being used to weigh finished half gallons of milk to the ounce, then the tolerances should be equal to at least one tenth of an ounce).
- Activities to be performed should the equipment be found reading out of tolerance.
- Requirements for protecting the equipment.
- Records to be maintained as "quality records" demonstrating compliance to defined requirements.

The Calibration List

The ISO standard requires that the organization "identify all inspection, measuring and test equipment that can affect product quality" (ISO 9001:1994

Section 4.11.2 b). It does not require that a list be maintained. However, many organizations have found it is very successful to create a list of the equipment that not only identifies the equipment but includes all the important aspects of the program including the unique identification number, frequency, tolerances, date done, date due, and so forth. This provides a useful tool for monitoring and maintaining the system. Many actually create this document and maintain it as a controlled work instruction. A word of caution: To avoid having to revise this document every time a calibration is performed, it may be more useful to define that the completion date and next due date are maintained on another more flexible document. Many systems actually maintain this information in the preventive maintenance database. As previously stated, this is an excellent means to track and trigger calibrations as they become due.

Identifying the Date Due

When defining the "next due date" be careful that the related procedures and work instructions clearly define how this date will be identified. Some identify a specific month and day, others identify it as the last day in the month that it is due. In other words, if scales are calibrated every 6 months and the last date was January 15, 2001. Should the date due be June 15, 2001? If it is identified in this manner on the calibration sticker, then don't waste your breath trying to justify to the auditor on June 20, 2001, why it is acceptable that the equipment is still in use. To save numerous headaches, it is recommended that the label reflect the month and the year only, such as June, 2001. Identifying the date in this manner makes accountability much clearer and easier to maintain. Ensure that related procedures and work instructions clearly define this requirement and be absolutely sure that the calibration service provider understands your defined requirements. Many nonconformances have been written because of this very issue. If the sticker states overdue on June 15, then the operator should be trained not to use this equipment on June 16. Remember, the system must clearly define what is required, then perform the task as required, and have records to prove that it is done.

Protecting the Equipment

Equipment should be used in an environment for which it is designed. Efforts should ensure that the environment is equivalent with the capability of the equipment, consistent with the required measurements, and that it does not affect its accuracy. For example, a laboratory balance designed for use in a controlled atmosphere situation would most likely be affected in a wet and/or dusty process area. The equipment must also be protected from tampering with the calibration adjustment or any other situation that may invalidate the calibration status.

The Employee's Role

Edward Link (1997) in *An ISO 9000 Pocket Guide for Every Employee* describes the employee's role when using critical equipment:

> When an employee is about to use an instrument to make a measurement that reflects or affects the quality of product, it is his/her responsibility to know that the label or record indicates that the calibration is still valid. Also should any event (like dropping a measurement device) occur that might cause the device to go out of calibration, it is the responsibility of anyone having knowledge of this event to take the steps necessary to assure that the device is calibrated again. (p. 46)

Equipment Found Outside of Defined Tolerances

Activities and responsibilities must be defined for evaluating product that has been tested on equipment that is found to have been measuring out of its defined tolerance range. The standard states that the organization must "assess and document the validity of previous inspection and test results when inspection, measuring or test equipment is found to be out of calibration" (ISO 9001:1994, 4.11 f). Product dating back to the previous acceptable calibration test must be evaluated to ensure its conformance status. Any product found to be nonconforming as a result of these activities should be addressed as defined in the system's procedures and work instructions for nonconforming product. Calibration-related procedures or work instructions should reference those documents.

This particular requirement can produce many headaches and subsequent system nonconformances if not carefully defined. When evaluating the test equipment and setting requirements, be sure to consider the effect that an out-of-tolerance situation would have on the product. The defined tolerances must be realistic. It would be a benefit to the system to distinguish between tolerances that require equipment repair from those that would directly effect the acceptance of a product. Many systems perform back-up validation checks that can be used to verify that the product was actually acceptable. This can save considerable time and effort from having to retrieve and evaluate months of product data. Be careful when defining requirements. Remember you must define what you must do and do it. Records must be maintained to prove it. Set realistic and useful parameters.

Records received from external service suppliers should be carefully reviewed to ensure that all out-of-tolerance situations are addressed internally. Many times this is overlooked. Frequently, the records will be filed without confirmation that the equipment tested within tolerance limits. Requirements for reviewing and signing these records before filing should be defined in related procedures or work instructions.

NIST or Known Standard Traceability

Quality records must be maintained for critical equipment that demonstrates that it was calibrated "against certified equipment having a known valid relationship to internationally or nationally recognized standards. Where no such standards exist, the basis used for calibration shall be documented" (ISO 9001:1994, 4.11.2 b). Equipment traceable to a National Institute of Standards and Technology (NIST) standard is the most common and most efficient method. This group was established by Congress to provide official measuring criteria and standards to ensure product reliability. NIST standards are considered the official measuring certification for inspection, measuring, and test equipment.

If a NIST standard is not available, then records must link to a known standard that ensures accuracy and repeatability of the equipment reading. Examples of a known standard may include a gas liquid chromatography (GLC), which is calibrated by a curve created according to manufacturer's instructions. This curve is used to calibrate the equipment to ensure that predefined "accuracy" is achieved and that results are consistent. A colorimeter is another example of equipment calibrated by a "known" standard rather than to an NIST standard. Standardized "color tiles" used to confirm the accuracy of the equipment are provided from the manufacturer. Requirements for creating, identifying, and maintaining the known standards must be defined in related procedures or work instructions.

Validation Versus Calibration

There is considerable confusion between the terms "validation" and "calibration." So much so that an explanation has earned its own section within this chapter. Generally, the term calibration in the world of ISO compliance refers to the activities performed as described in this text to confirm that inspection, measuring, and test equipment provides accurate and reliable results. Calibration refers to complete compliance with these requirements, including the maintenance of appropriate quality records that provide traceability to NIST or a known standard. Some confusion sets in because this term had been used for many years in the industry to basically refer to checking the equipment, adjusting, and so forth. Some testing equipment even has a switch or push button labeled "calibration." When it is engaged, the machine checks and resets itself. It is important that associates responsible for calibration requirements understand the meaning of the term as it relates to the ISO-compliant system.

Further confusion arises with the use of the term "validate." Many use this term to perform routine checks on equipment to ensure that it is working properly and giving accurate results. An example of this may be when a work instruction requires that before a scale is used it is confirmed that it is reading at 0.0 grams. In reality, probably the only difference in actual performance is

CONTROL OF INSPECTION, MEASURING, AND TEST EQUIPMENT   **145**

the amount and type of records that are maintained. It would be extremely burdensome to have to record all the necessary data, including traceability, every time one performed a routine weight on an analytical balance. Validation results can be very useful as references. They provide important historical data on equipment should a calibration check be found reading out of tolerance.

How the terms are used is up to system management. Whatever terms are used and how they apply should be clearly defined in appropriate procedures or work instructions with quality records maintained to demonstrate that associates responsible for these activities are trained.

Quality Records

Records must be maintained to demonstrate compliance to defined requirements. Records for calibration are very comprehensive and should include the following information:

- The unique number (identification) of the equipment.
- Date calibrated and date next calibration due (confirms defined frequency).
- Required tolerances (how accurate must the reading be).
- Actual reading "as found" (confirms instrument tolerance status).
- Reading "as left" (confirms tolerance status at completion of the calibration exercise).
- Evidence that the validity of results of product tested on equipment that is found reading out of tolerance during its calibration has been evaluated with appropriate actions taken as a result of this evaluation.
- The unique identification of the test equipment used to perform the calibration. (Records must be traceable to national, international, or a known standard.)
- Identification of the person (s) performing the test.
  - If performed internally, then this will provide the trail to confirm that this individual has met the training criteria (qualification) for this activity.
  - If an external service company is being used, then this information (company name) should provide the trail to ensure an approved supplier has been used and that it is meeting all the defined (communicated) requirements.

Manufacturing manuals and other similar type documents that are identified as the source of required information must be established and maintained as external documents. Requirements for external documents can be complex and are addressed in the Document and Data Control section of this text.

External Calibration Service Suppliers

It is very common to contract an external calibration service company to perform calibration activities. This can be expensive, but in many instances it may be much more cost effective then purchasing and maintaining the "test" equipment internally. The external supplier of calibration services must be identified, approved, and evaluated as an approved supplier in a manner that meets the system's defined purchasing requirements.

The purchasing document, which may be a purchase order or contract, must clearly identify the product ordered. The product is the service of performing "calibration". The purchasing document must define all the requirements, including referencing required records, specifications, procedures, and the like.

Some organizations define the requirements in the contract with the purchase order referencing the contract. Others may define the requirements in a procedure or work instruction, which is then referenced in the purchase order with a controlled copy provided to the supplier. Another option is to reference the controlled procedure or work instruction in the purchase order and require the supplier to review this document on-site prior to performing the services. Remember there must be proof (initial/date) that the service representative reviewed the required procedures or work instructions.

It is absolutely essential that all records provided by the external service supplier define the necessary proof of compliance. Review the criteria and records required for compliance with your supplier. An individual should be assigned the responsibility for working with this supplier and ensuring that activities and records meet the organization's defined requirements.

Revising the Requirements

Over time, it may be determined that the defined calibration requirements are either too stringent or not stringent enough. It is important that the reasoning and appropriate data related to program revisions for such aspects as frequencies and tolerances are clearly documented. It is your system to define, but evidence and records prove invaluable both from an auditing perspective and most importantly from a process improvement viewpoint.

Management's Role

Edward Link (1997) in *An ISO 9000 Pocket Guide for Every Employee* provides an excellent description of management's role within an effective and compliant quality management system:

> Management must encourage and provide for a calibration program that includes all considerations found in [ISO 9001:1994] element 4.11. Not doing so could directly compromise the quality of the products. There is also an indirect threat to the quality of products when calibration of instruments used for process

CONTROL OF INSPECTION, MEASURING, AND TEST EQUIPMENT 147

control measurements is not appropriate. Management must assure that appropriate expertise is available to understand and fulfill your company's measurement needs. (p. 45)

Frequently Identified Nonconformances

- Not all critical equipment has been identified with records to confirm compliance.
- Review of records from the external service supplier indicates that in several instances equipment had been found out of the tolerance ranges; however, there was no evidence that the validity of product tested with this equipment since the last acceptable calibration has been investigated.
- Several pieces of equipment were being used within the process to demonstrate compliance to specified requirements with past due dates on the calibration stickers. No records were available to confirm whether tests had actually been performed on time.
- Several pieces of equipment were being used within the process to demonstrate compliance to specified requirements with past due dates on the calibration stickers. However, review of records indicated that the equipment had recently been calibrated and not overdue.
- Status of calibration for critical equipment was not readily available to associates using this equipment in the processing and testing areas.
- Through the interview process, it could not be confirmed that associates understood their responsibilities as defined in related procedures and work instructions for performing validation versus calibration activities.
- Records provided by the external calibration service company did not provide identification traceable to a national standard or a known standard.
- Serial numbers listed on the master calibration list did not match the serial numbers on the equipment actually being calibrated.
- Some scales and thermometers used to demonstrate compliance to critical test parameters were not included in the calibration system.
- Scales used on the process blend area had two different conflicting calibration labels.
- Lot numbers that provide NIST traceability for pH buffers were not recorded on the calibration log.
- In some instances, pH buffer was being transferred into dispensing bottles. Bottles did not include lot number identification that would provide NIST traceability.
- Pentone color standards were being used to confirm color; however, evidence could not confirm that these standards were being controlled.

9.4 CORRECTIVE AND PREVENTIVE ACTION

An effective corrective and preventive action process is absolutely essential to the quality management system. Corrective actions focus on "existing" nonconformances. Preventive actions focus on "potential" nonconformances and addressing problems before they happen. Corrective and preventive actions "taken to eliminate the causes of actual or potential nonconformities [must] be to a degree appropriate to the magnitude of problems and commensurate with the risks encountered" (ISO 9001:1994 4.14.1).

A compliant preventive action program evaluates "appropriate sources of information such as processes and work operations which affect product quality to detect, analyze and eliminate potential causes of nonconformities" (ISO 9001:1994 4.14.3 a). A nonconformance or nonconforming situation is an activity that is not being performed as defined by the quality system.

Thoughts on Corrective and Preventive Action Process

The corrective/preventative action process is really the mechanism for evolving your system. It's the driver for continuous improvement. What was essential to the success of the corrective/preventative action process was that the procedure was simple and accessible to everyone, system users and customers. Any time someone had a complaint, or found an error in the system, or discovered a better way of doing something, they simply documented it on a CAR/PAR form. The process did not allow the idea to get lost; it had to be acted on and followed up on for effectiveness. With the system working properly, we were never at a loss for improvement ideas.
—Victor V. Margiotta, Director of Quality, SOBE

This is the reason most people want a formalized quality management system, to fix existing problems and prevent future problems.
—Bill Lockwood, Package Quality Manager, Hiram Walker & Sons, Ltd.

In my opinion the most useful system/benefit that has been experienced as a result of certification is the corrective and preventive action system. I am finding that having a channel for the identification and execution of corrective actions is driving the changes that are necessary to improve our business systems from within and to improve our ability to meet our customer's expectations. It is a tremendous benefit having a structured program whereby actual and potential problems can be documented and addressed. The structure of this system allows for a systematic and organized approach to issues and also requires that the root causes are addressed completely. The best things about developing a corrective/preventive action system is having a formal channel to address internal issues and customer complaints, to address potential issues before they get out of hand and to allow all personnel the opportunity to change the systems that affect his or her work. Every issue is handled through the same channels regardless of the source. Our system calls for the weekly meeting of a cross section of the company's management to review the week's corrective/preventive action requests and customer complaints. This system has opened channels

of communication that have not previously existed and provided a measure of accountability to see changes through. It also provides a forum for feedback on the corrective/preventive action system on a weekly basis. I'm not sure that we would have the same success at continuous improvement without this program. The corrective/preventive action system requires accountability to the quality system. This is a powerful tool to drive change.
—Russ Marchiando, Quality Systems Coordinator, Wixon Fontarome

Robert Peach (1997) in *The ISO 9000 Handbook, Third Edition*, provides an excellent description on exactly what this process means to the entire system and some of the problems related to its management:

> Often the weakest part of quality systems, corrective action loops [activities to address the findings] are frequently designed only to address the immediate problem while failing to act to avoid its recurrence. Another common problem is that they often deal only with matters of processes, products, or services while overlooking the system. The standard requires a rigorous examination of all the quality data and records to detect and remove all potential as well as actual causes of nonconformance. This is proactive quality, not reactive.

Peach goes on to state that it is by defined requirements that the ISO standard

> reinforces the primary intent of the standard, which is preventing nonconformity at all stages. . . . Investigation of nonconformance has to operate on three levels: "investigation of the causes of nonconformities related to product, process, and quality system." Any one of the three can be the cause of a nonconformance and therefore require corrective action. Companies with systems that only apply corrective actions to product issues must extend that system to cover the other two sides [process issues and system issues] of the conformance triangle. (p. 139)

Food Guidelines (1995) states:

> Once a problem has been identified, the system must be investigated to find what went wrong and stop it from happening again. . . . It is not sufficient just to correct a problem, which has been found although this is always of prime importance. Most problems occur either because the system has not been followed or because there is a loophole in the system. Corrective action must also involve finding the root cause of the problem, correcting it and giving instruction so that it will not happen again. This is the area where the greatest cost saving can be made, and is the main point of having the documented system. (pp. 37–38)

The Quality Records

Records maintained in compliance to the system's defined quality record process must demonstrate compliance to the following requirements:

- Activities include the "investigation of the cause of nonconformities relating to product, process and system issues" (ISO 9001:1994 4.14.2 b).
- "The effective handling of customer complaints and reports of product nonconformities" (ISO 9001:1994 4.14.2 a).
- Root cause analysis and identification of what is required to correct or prevent the nonconformance.
- Documented follow-up activities that confirm the actions taken were effective.
- Relevant information regarding the preventive action process "submitted for management review" (ISO 9001:1994 4.14.3 d).
- Activities as a result of corrective and preventive actions include the implementation and recording of any changes to documents.

The corrective and preventive action records should be maintained for at least 3 years to provide sufficient evidence of the maintenance and effectiveness of system activities.

These are the requirements, but what do they really mean? The corrective and preventive action process is the heart of the system. It provides a structured means for improvement through identifying opportunities that, when addressed, strengthen the system as a whole.

It is recommended that an associate with responsibility and authority within the system be assigned as the manager or coordinator of the corrective–preventive action process. Many times this is actually assigned to the management representative who will either manage it or assign the responsibility to another position to perform under his or her direction.

In the *ISO 9000 Handbook, Third Edition*, Peach (1997) compares the related corrective action activities to

the well-known "plan-do-check-set" cycle

- plan the action to be taken after identifying and evaluating possible solutions,
- take the action,
- confirm that the action was effective
- continue to monitor results to ensure that the problem has been solved (p. 230)

When identifying a corrective action, it is important to document it clearly and ensure that it is a statement of fact. It should not be used as a means to apply discipline or as a process to air gripes. The commitment date should be directly related to the potential risk of the situation. The process itself should clearly define the criteria for identifying and issuing corrective and/or preventive actions. Be careful not to make the system more than it is.

Corrective and preventive actions should be applied to product, process, and system issues. This includes customer compliance and nonconforming product. However, be cautious when including customer complaints and nonconforming product in this process. It is much more effective to monitor and record these on an individual basis through a process designed especially for that purpose. This information should then be monitored and evaluated for trends and situations. Then, where appropriate, it should be documented within the corrective–preventive action process. For example, one complaint on a loose bottle cap may be addressed on an individual basis with the customer, however, if several complaints on the same issue are received, then this may require a formal corrective action. It could very likely overburden the process if a responsible manager had to respond formally through the corrective action–preventive action process for every individual instance.

Most food companies already have a well-defined means for handling customer complaints and nonconforming product. Most often these programs are incorporated into the system, then monitored for trends. Concerns may then be investigated through the structured format provided by the formal corrective–preventive action process. The process is at its peak when these trends and concerns are formalized through the corrective–preventive action process.

In the food industry, the corrective–preventive action process should also be applied to nonconformances resulting from sanitation, GMP (Good Manufacturing Practices), and pest control audits. In addition, the use of hazard risk analysis techniques should be applied to preventive actions, where appropriate.

Concerns Regarding the Corrective–Preventive Action Process

At the onset we were setting ourselves up for failure and struggled mightily by attempting to address every complaint through our corrective/preventive action system. It took awhile to realize that everything that goes wrong in a company is not necessarily worthy of a corrective or preventive action. It was also difficult changing a culture where the documentation of problems and solutions were nearly nonexistent. By requiring the documentation of a formal root cause analysis, its corrective action and follow up investigation, we received significant resistance from people claiming that this process was "just more paperwork" to deal with. Even as our system matures it is difficult to keep people from taking shortcuts on the paperwork. Finally, another problem that we have is the use of the system to air departmental gripes. Instead of working out minor problems of communication outside the system, some people are content to use the system to solve all of their problems for them.

—Anonymous

One of the most difficult aspects of maintaining the system after certification has been getting true corrective action and real change. All too often the quick fix at the last minute was used. There wasn't the resolve or mandate to institute truly effective corrective action. Those who were charged with maintaining the system were always on edge walking a fine line between making sure the problem appeared fixed and knowing that more could and should have been done.

—Anonymous

As the system is being implemented, it is critical not to become too bogged down in the corrective/preventive action process. Frequent periodic audits of the status of corrective actions will help prevent the death spiral.

—Jim Murphy, Manager of Design Process and Validation,
The Dannon Company

Documenting the Corrective Action

There are many different types of forms used to document corrective and preventive actions. It is recommended that a form be created that addresses the following corrective–preventive action activity chain. Identify the nonconforming situation, which may also be referred to as either the "finding" or "noncompliance."

- Identify the source of the noncompliance (i.e., internal audits, management review customer complaints, product nonconformance, employee feedback, failed effectiveness review, external audit, etc.).
- Review and assign the corrective action to the responsible department manager.
  - Responsible manager may proceed or assign it to a designee to work on the issue under his or her direction.
- Manager reviews the noncompliance and performs whatever is required to identify the root cause then documents the root cause and proposed corrective action.
- Identify a realistic commitment date to have the corrective action completed based on the potential risk of the noncompliance. Every effort must be made to address the corrective action in a timely manner.
- Complete the corrective action and document the action taken and the date completed.
- The corrective–preventive action manager or coordinator assigns a date to evaluate the effectiveness of the action taken. The elapsed time period should provide the opportunity for significant evidence/activities to occur such that the results of the actions taken can be effectively evaluated.
- Effectiveness is evaluated and documented, as appropriate.
  - Confirmed effectiveness completes the corrective action.
  - Unconfirmed effectiveness should be documented. The corrective/preventive action form (CAR/PAR) would then be rerouted, as appropriate, depending on the defined process, which may be to either reissue the current form or initiate a new one.

A sample of a corrective action request and preventive action request (CAR/PAR) is presented below. Keep in mind that there are many different versions that can be developed, but basically this is the key information that should be recorded.

CORRECTIVE AND PREVENTIVE ACTION

Corrective / Preventive Action Request (CAR/PAR)

CAR/PAR # _____
☐ Corrective   ☐ Preventive

Originator _____   *Date* _____

Identify the source of the nonconformance
- ☐ Customer Complaints
- ☐ Product Nonconformance
- ☐ Not Following Procedures
- ☐ Not Defined in Procedures
- ☐ Management Review
- ☐ Supplier Nonconformance
- ☐ Internal Audit
- ☐ External Audit
- ☐ Follow-up Audit
- ☐ Rejected CAR/PAR Approval
- ☐ Trend Analysis
- ☐ Nonconforming Supplies

Reference documentation attached and identified (yes/no)

Describe of the nonconforming situation:

To be completed by the CAR/PAR coordinator or designee:

Assigned to (identify responsible manager): _____
Date assigned: _____

Assigned manager or designee completes the following:

Root Cause Analysis:
(identify the proposed cause of the nonconformance)

Prosposed Action to be completed
Completion Date: _____ _____
     ⟨*Proposed*⟩   ⟨*Actual*⟩

The CAR/PAR coordinator or designee completes the following:

Proposed effectiveness review Date: _____   Actual Date: _____

Provide a brief description of the results of the evaluation for effectiveness:

(Describe evidence of effectiveness review. Clearly define reasons and linkage to references should CAR/PAR effectiveness not be approved resulting in further activities or newly raised CAR/PARs.)

Effectiveness accepted (yes/no)
Describe any additional requirements (i.e., reissue, etc.)
CAR/PAR completed: _____   By: _____

One of the important aspects of the corrective/preventive action process is that it focus on identifying and solving existing and potential problems in a timely manner with a strong focus on preventing reoccurrence. The process requires that the root cause is identified, investigated, and documented.

The root cause is addressed through the actions taken. A significant benefit of the process is realized when results of these actions are evaluated for effectiveness. In doing so, it may be determined that the action was not effective. The original root cause analysis, although the obvious or most likely cause, was not the complete reason for the nonconforming situation. This forces further review of the situation, identifying and addressing root causes until the actions taken are confirmed effective. An industry professional who must remain anonymous due to company policy provided the profound description that the corrective/preventive action process "is the To Do List that doesn't go away."

Some organizations choose to have a combination of many different corrective action processes to address nonconforming situations. These could be for such areas as nonconforming product, supplier nonconformances, and internal audit nonconformances. Other organizations prefer one process with the sources identified on the corrective/preventive form itself. Although multiple processes can be successful, a single process may be easier and more efficient to document, maintain, and monitor. It is common during the implementation stage to possibly have two or more distinct processes for handling nonconformances; however, as the system matures, management usually decides to combine the reporting and documenting format into one process. Through experience, it has been observed that associates feel more comfortable having to learn one process, using one form, and having one specific means to address existing and potential nonconformances. Keep in mind

CORRECTIVE AND PREVENTIVE ACTION 155

though, that this central system should not be expected to address every single product nonconformance or customer compliant. These should still be monitored for trends and concerns and then, as appropriate, documented and managed through the formal corrective–preventive action.

Identifying the Nonconforming Situation ("Finding")

The first step of a documented corrective/preventive action is to describe the situation, (the nonconformance). It is important that this is a description of the situation—a statement of fact—and not an explanation of what went wrong or how to fix it. Some examples follow:

1. Contract review documents are not being signed and dated as required by procedure CON-01 Rev. 2.

This example is a statement of fact. It identifies the nonconformance and the fact that it is an activity not being performed as defined in a system document.

2. Training procedures and work instructions do not define the requirements for associates performing blending and packaging activities.
3. Training procedures and work instructions must be written to define the requirements for training blending and packaging associates.

The difference between examples 2 and 3 is that example 2 states the fact whereas example 3 directs how to fix it. The how to fix it should be identified by the person responsible for fixing the situation. More times then not, the person identifying the situation doesn't know the facts nor has the knowledge of the situation to identify the root cause. Keep in mind, even if he or she does, by dictating the fix, this person is accepting ownership of the nonconformance. Ownership belongs to the responsible department manager. Generally, systems mature to be more proactive and effective long term, when identifying the root cause and the action to be taken is assigned to the person responsible for addressing the finding.

4. Incoming inspection log is not identified on the quality record listing.

This could be classified as an improperly written finding because it gives direction. It is describing the fix rather than statement of fact. Item 5 below states the facts.

5. Review of the documentation and activity for performing incoming inspection did not provide evidence that records are being maintained as required by procedure QA-01 Rev 0.
6. Although procedure PM-02 Rev 2 states that suitable maintenance will be performed to ensure process capability through a preventive

maintenance program, at the time of this evaluation there was no program defined and implemented.

Example 6 states the fact. It references the system's defined requirements. Compare this to example 7, which is "assuming" or "directing" that there should be a preventive maintenance program. As will be discussed in regard to process control. ISO 9001:1994 does not specifically require a preventive maintenance program. Findings should not be written assuming the answer or giving direction.

7. There is no evidence of a preventive maintenance program.

Example 6 references where it is stated in the system's documents. This dictates that there should be a preventive maintenance program. If documentation does not reference how "suitable maintenance would be performed to ensure process capability," (ISO 9001:1994 Section 4.9 g) then the following statement of fact would be an appropriate finding.

8. Documentation does not define the manner in which the system will perform suitable maintenance to ensure process capability.

Writing findings can be very tricky. The system will not disintegrate if the wording isn't exactly as it should be.

Example 9 is a very good example of leading the fix.

9. Numerous cartons of product are being damaged and lost because the warehouse roof is old and needs replacement.

The decision as to "why the roof is leaking and damaging the product" should be left to the person responsible for addressing the finding. Number 10 is closer to being a statement of fact.

10. Numerous cartons of product are being damaged and loss due to leaks in the warehouse roof.

Actually reviewing the findings and assigning the responsibility should be part of the corrective/preventive action coordinator's function. Some systems identify a corrective/preventive action team that reviews the documented nonconformances and then assigns the corrective/preventive action to the responsible manager. This is usually very successful and effective. The team also monitors timeliness and assignment of resources. Results and any concerns related to the corrective and preventive action process as identified by the team are reported at the management review meetings. It is at the management review meetings that executive management can assign resources and address concerns, as appropriate.

Root Cause Analysis

The root cause should be identified by the person or group responsible for addressing the situation. Some systems may encourage internal auditors to "talk over the solution" with the area associates. This can be helpful, but the auditors must be careful not to offer the fix. The root cause analysis is what the "fixer" feels is the cause of the situation. From this decision, the corrective action is identified. Auditors that dictate the root cause and/or the fix (action to be taken) will be assuming ownership of the situation.

The Corrective Action

The corrective action is the action being proposed to address the root cause. Addressing the root cause is the reasoning process applied to correct or prevent the nonconforming situation. The corrective action should be identified by the responsible manager or designee. The parties whose responsibility it is to fix the situation, to identify the root cause, should be the ones to identify what will be done to correct or prevent the situation.

Identifying the Corrective Action Date

The ISO standard states that the corrective action should be completed within a time period directly related to the potential risk of the situation. A commitment should be made for a realistic, practical, and good-faith date. Many times this "commitment" becomes an issue because responsible associates hesitate to choose a date, feeling that there will be negative retributions should the date not be achieved. This system must not be presented as a "fear" system. Issues and commitments must be looked at through a positive eye toward improvement opportunities. Opportunities are a means to strengthen the system through the identification of existing and potential nonconforming situations. Once a commitment for completing the action is documented, then the corrective/preventive action coordinator or team should evaluate the time frame to ensure it is adequate in relation to the potential risk of the situation. Any concerns should be discussed with the responsible manager.

Requirements for approving and documenting extension dates for completing corrective actions should be clearly defined. It is recommended that the request be presented in writing with a clear description of why the extension is required and identifying a new date for completion. Approved extension dates should be documented as such. Dates that are just not achieved are considered as "overdue". Let it be emphasized that tracking the number of open and closed (completed) corrective and preventive actions can communicate a negative message. This process should not be considered a race to open and close actions as quickly as possible, but as a useful process to address opportunities for improvement effectively and efficiently. It is important for the responsible manager to commit to a realistic date based on the potential risk of the situation and then to complete the action by that date. Monitoring the process with this focus provides data on the timeliness of actions taken rather than the number of corrective/preventive action reports closed. Being in too much of a hurry to close actions will decrease the opportunity to effectively address them. The system should be monitored by the timeliness/on-time completion of the corrective/preventive action activities, absolutely not on the number that are closed. Justifiable (approved) completion date extensions should not be tracked as "late."

"Closed" or "Closed"

Depending on how the corrective/preventive action process is defined, the corrective action report may be "closed" once the corrective action has been completed. Some systems, however, define "closed" as when the effectiveness has been verified. This may be only a discussion on semantics. It is important that the process be defined in a manner that is best for the overall operation. From experience, it seems that most matured systems tend to define the CA/PA as closed once the action itself is completed. Evaluation for effectiveness is performed as its own activity. Should effectiveness not be confirmed, then a new CA/PA is issued with a copy of the previous report attached for historical information. Documenting and tracking this information on the CA/PA action log provides information on those that must be reissued. This is not done to find fault but to provide information and background to promote an overall effective corrective and preventive action process.

Evaluating the Effectiveness

As mentioned previously, a time period should be identified between the completion of the corrective action and the evaluation for its effectiveness that provides significant evidence or opportunity for the situation to reoccur. For example, if the roof is leaking, the corrective action may be to patch the roof; however, the effectiveness cannot be truly evaluated until it actually rains. Another example would be if the finding was that "forms in the blending area were not being completed as required by documented procedures and/or work instructions." Initially, the responsible manager may identify the root cause to be inadequate or insufficient training. The commitment for training is identified and performed. Depending on how often the forms are filled out, in this case weekly, the corrective/preventive action coordinator decides to allow 8 weeks from the time of the training to evaluate its effectiveness. After 8 weeks this review determines that the situation is improved, but still not totally in compliance. Further evaluation of the root cause identifies the fact that the form itself is not designed in a useful manner to promote compliance to defined requirements. The activity is then reevaluated with a new root cause identified as the design and content of the form. This is the major advantage of the structure and discipline of an ISO-compliant system. It requires that the

CORRECTIVE AND PREVENTIVE ACTION 159

effectiveness of the action taken is confirmed. Many times "Band-aid" fixes are identified and applied in good faith, truly believing it is the correct fix. The initial appearance is that the situation has been corrected; however, the true root cause has not been addressed thus the situation reoccurs.

In Review

Effectiveness is not verifying that the corrective action is done. Completing the corrective action is the responsibility of the person assigned the corrective action. That person or group must take ownership for their commitment. When the corrective action is completed, the responsible individual should document the completion and return the form to the coordinator. Verification for effectiveness is just that, verifying that the corrective action was effective. The time period necessary to determine the effectiveness should be realistic, providing enough evidence to truly evaluate if the fix is really fixed. It should not be a "pressure" date created in an attempt to complete a corrective action report.

Documenting the Effectiveness

The corrective/preventive action report should contain a clear and precise description of what was evaluated to determine if the actions were effective. This should not be a restatement of the finding or actions. For example, if the finding is that Bob and Jane's training records weren't up to date, a description of the effectiveness evaluation may state: "Reviewed a sampling of Bob's, Jane's, and 10 other randomly chosen associates in the department and found all training records in compliance."

True effectiveness could not be confirmed if only Bob's and Jane's file were checked. The following statement would not be appropriate: "Bob's and Jane's training files were confirmed present." Remember, the goal is to address the activities as a whole, just fixing the two noted files would not be a clear indication as to the total compliance of the situation.

Another example may occur during an audit: through the interview process some associates did not know the quality policy and the role that they played in achieving it. "The root cause was identified as insufficient training, which according to the responsible manager, was addressed through a special refresher training session. To verify the effectiveness of the action, a random sample of the associates in the area should be interviewed. The effectiveness may be documented as follows: confirmed training had been completed by review of records. Sampled approximately 10 associates in the area and confirmed that they were knowledgeable on the quality policy, his or her role in achieving it, and where a controlled copy of the policy was located."

Buried Paperwork

A good way to ensure that corrective/preventive action reports that need attention do not get lost in the paperwork shuffle is to print the forms on bright-colored stationery. Bright red or kelly green will stand out in a pile of white papers. One company changes the color every 6 months to help monitor time frames.

The Corrective/Preventive Action Log

Many systems find the best method to track the status of corrective and preventive actions is to use a log or chart that lists the corrective action by number, the responsible person, the projected date for completion, actual completion date, date to be evaluated for effectiveness, and date effectiveness is confirmed. Some systems create a code that readily distinguishes between preventive and corrective actions. This type format will not only provide the opportunity to track individual and completed items, but it makes an excellent information source for reporting on the status of the corrective/preventive action process at the management review meeting.

Food Guidelines (1995) describes the role of statistical analysis:

> Some statistical analysis should be included. There is little value in spending money and resources to investigate simple one off errors. There must be a method of identifying repetitive problems and differentiating between "honest mistakes" and disasters. Control charts can often prove useful in assisting early identification of trends in continuous processes. Plotting results on charts gives a more obvious indication. (p. 38)

Thoughts on Preventive Action

> One of the most difficult aspects of maintaining the system has been "preventive action." While we continually do things that are preventive in nature, we do not do a good enough job of documenting them. I believe this has more to do with our midwestern agrarian culture of constantly being on the lookout for things to fix and taking care of them as a matter of fact. I truly feel that many folks think of preventive action as being part of their everyday responsibility and do not appreciate it as a noteworthy event. As such, they do not always formally document the activities.
> —Dana Crowley, Production Manager, Danisco Sweeteners, Inc.

Food Guidelines (1995) discusses long-term improvement projects:

> There should also be provision for longer term improvement projects. Whilst the system may be basically sound, there will always be areas in which it can be improved. These may be identified through management review and may relate to non-production systems. Examples may include improvements to the training system or company organization. The system should be considered flexible and constantly needing [to be] reviewed if the company is to maintain its competitive edge. But all changes must be reflected in the system to ensure they are properly communicated and understood. (p. 38)

CORRECTIVE AND PREVENTIVE ACTION   161

Is It a Preventive or a Corrective Action?

In some instances, the difference between a corrective and a preventive action may be very evident; however, overall, it might be difficult to differentiate one from the other. First of all, it is important to realize that the system is not going to disintegrate nor will the ISO police be called if a preventive action is labeled a corrective action or visa versa. It is important to understand the basic concept that corrective actions address nonconforming situations whereas preventive actions address *potential* nonconforming situations. In many instances, there is a fine line between an existing noncompliance or a potential one. Labeling is not as important as the fact that the process is being applied. If the roof is leaking, patching the leaks is considered a corrective action. If the roof is old and the decision is made to replace the roof before it starts to leak, then a preventive action has been taken. Think about some of the following examples. Are they preventive actions or corrective actions?

- After being asked to photograph your nephew's wedding, you decide to have your camera checked out to ensure it is operating properly.
- The gas tank is running low, you decide to stop at the next gas station just in case there isn't another one within the next 30 miles or so.
- My car ran out of gas 10 miles from the nearest gas station, one hour before my plane was due to leave.
- The coding on the finished carton is fading. Although it can be read, it is difficult.
- The code date on the finished product carton has become illegible. The printer requires a new roller and ink jet cartridge. There are no spares on-site and the replacement may take 5 days for delivery.
- Although internal audits are being performed on time, the trained auditor force has decreased approximately 50% in the past year. Because of this situation, it was decided to train 10 new auditors within the next 6 months.
- Fifty percent of the audits have been more than one month late. Root cause analysis determined that the cause is most likely due to the decrease in available trained auditors. The decision is made to train 10 more auditors which will bring the trained auditor pool back to its original number.
- Although the maximum allowable storage temperature for the cold room is 40°F, maintenance notices that instead of its usual reading of 33°F it is now running at approximately 38°F. The decision is made to rent an auxiliary unit and immediately overall the present unit.
- You change the oil in your car every 4000 miles.
- You have driven 20,000 miles and never changed the oil in your car. You have decided to run the engine until the existing oil is gone.

- A new process line is going to be operational in 30 days. Procedures and work instructions, the absence of which would affect quality, are being prepared and incorporated into the document control function. Plans are in place to train operators on the documentation prior to startup of the process.
- A new operation has been operating for approximately 30 days. Procedures and work instructions have not been written. Operator training has been informally performed by the manufacturers; however, the training was not documented. To date only a limited amount of product produced has met specifications.

Long-Term Corrective Actions?

Some associates have difficulty between long-term corrective actions and preventive actions. In addition to patching the leaks, a decision is made to approve an entire new roof to prevent more leaks. Is the new roof a corrective action or a preventive action? What is described here is considered a gray area by many. The important issue is that the situation is corrected and hopefully the opportunities for it to return minimized. How it is classified is not necessarily relevant. It is recommended that the corrective/preventive action procedures and work instructions define these classifications, as appropriate. Some systems actually document short-term and long-term actions where appropriate, but this may be too complex over time. It is recommended that each issue be discussed and handled in the best manner to be useful, and promote continual improvement opportunities within the system.

Reporting to the Management Review Meeting

Review of the status and activities regarding the corrective and preventive action process should be presented at the management review meetings. The status of the process may be simply reporting on the timeliness in addressing the requirements, with any issues and resource requirements discussed and assigned as appropriate. This is one of many instances when the standard is inherently requiring management to be involved with the system. Robert Peach (1997) in *ISO 9000 Handbook Third, Edition*, describes that "the standard is seeking to ensure that management becomes fully engaged in operating the system" (p. 139).

Ongoing Preventive Actions

Many situations inherent within the system may be defined as ongoing activities performed in a manner to prevent nonconforming situations. Examples of these may include: HACCP (Hazard Analysis Critical Control Point), developmental training, GMP/sanitation audits, team building, and preventive maintenance. These type programs may be discussed at the management review meetings as an identified agenda item titled "preventive actions." Summaries or attachments providing proof and actions on these discussions should be maintained as part of the management review record. Discussing corrective and preventive actions with management is further evidence for the proactive nature of the ISO-compliant system.

More Thoughts on the Corrective/Preventive Action

> I feel there have been numerous process/system benefits of obtaining ISO 9001 certification that have helped move our quality system to the next level. However, I think the single biggest benefit has been the institution of a true corrective and preventative action based system. If properly structured and utilized its potential is unlimited as a vehicle to drive continuous improvement throughout the organization. It provides a structure to address issues at any level of the company and in every department within the organization. It provides a structure not only for identification and documentation of current or potential problems but it also demands a root cause determination be done and a commitment by the person(s) responsible for proposing what corrective action is to be instituted within a targeted time frame. The progress being made within this type of structured system can easily be monitored and periodic reports generated for distribution to all levels of management. The unlimited inputs into such corrective/preventative action make this system a very dynamic tool for organizational and quality improvement. The primary inputs to our corrective/preventative action process incorporate customer complaints (both external and internal), product returns, internal quality audits, customer audits of our facility, internal trend analysis, sanitation/GMP inspection results, safety committee inspection results, HACCP, and SPC system results and regulatory agency audits.
> —Tim Sonntag, VP Quality Assurance & Technical, Wixon Fontarome

In Conclusion

The corrective and preventive action process is a very critical aspect of every quality management system. It is imperative that this process be well defined and maintained in an aggressive and effective manner to achieve the highest degree of value and usefulness of the compliant system. This process will most likely go through many phases as the system itself matures. It is difficult to always know what should be recorded and what shouldn't. This can be achieved through effective management and basically living the learning curve. This is where the corrective/preventive action team can be very useful.

Edward Link (1997) in *An ISO 9000 Pocket Guide for Every Employee* provides an excellent conclusion to these discussions:

> Corrective action can be thought of as first aid for manufacturing and service problems. What is really needed is a greater emphasis on being proactive or the use of more preventive action. The most important characteristic of corrective and preventive action systems is that they be closed loop. Good closure means

that the problem has been thoroughly analyzed, the solution well thought out and implemented, and the effectiveness verified as required by ISO 9000.... Preventive action results in the reduction of variation. As the variation is reduced, the likelihood of or the potential for the occurrence of the addressed nonconformity is reduced.... The success of your company depends on the strength of your Corrective and Preventive Action system. Barriers get removed and problems get solved when management demonstrates strong support for the problem solving efforts occurring within the Corrective and Preventive action system. (pp. 54–57)

Frequently Identified Noncompliances

- It was not evident from review of the records that the effectiveness of the actions taken were reviewed and documented.
- Review of the corrective and preventive action reports did not provide evidence that actions were being performed in a timely manner based on the potential risk of the situation.
- Review of the records indicated that a significant number (at least 50%) of corrective actions were not being completed on time.
- The corrective action reports did not provide the complete record such as date to be completed and root cause analysis.
- It was stated that corrective action reports are monitored for timeliness; however, review of the reports indicated actual completion dates were not recorded.
- Review of a sampling of management review minutes did not always provide proof that the corrective and preventive action processes were being evaluated during these meetings as required by procedure MR-02 Rev 3.
- Records did not provide evidence that product, process, and system issues were being addressed through the corrective and preventive action process.

9.5 INTERNAL QUALITY AUDITS

The internal quality audit is one of the key activities described as the "heart" of the system designed to focus on confirming effectiveness and fine tuning the system. A well-defined and effective audit program is essential to the growth and maturity of the quality management system. The complete internal quality audit process is quite complex. It is essential that associates at all levels of the quality management system understand the purpose of these audits. These audits are not meant to find every possible noncompliance but to provide a snapshot audit with an extra set of independent eyes evaluating the process. It is essential that management and associates maintain their areas in compliance, addressing nonconformance issues as they arise, not waiting for the auditor to find them.

It is equally important that the audit be welcomed as a positive means of evaluation. The old saying "one can't see the forest for the trees" holds very true when applied to the management of individual areas within the system. Many times those responsible are just too close to the situation and can't see what independent eyes can see. The audit is only as effective as the auditor and the auditee team combined. It is a fundamental part of every successful quality management program.

Thoughts on Internal Quality Audit Process

Establishment and maintenance of a thorough, yet efficient internal audit program is the single most valuable tool an ISO certified company has to help make sure that [it is] doing what it states in documentation. These audits pinpoint areas that have fallen or could eventually fall out of compliance. They provide the framework via the corrective action process to assure [the system] remain ISO compliant.
—Tim Sonntag, VP Quality Assurance & Technical, Wixon Fontarome

The internal audit process was one of our biggest challenges to manage. It was even more time and resource consuming than writing our procedures! We did not have the luxury of having dedicated ISO auditors. We had to schedule employees' time away from their normal duties to perform internal audits. The way we managed the problem of resources was to get several people from each department trained as Internal Auditors. This allowed us to rotate auditors throughout the year based on an audit schedule. The benefits of this approach were many. First, everyone could find the time to perform an occasional audit. It fact, we usually looked forward to the change of pace from our normal duties. Second, auditors were assigned to audit a different department from the one in which they worked. The advantage to this was the auditor usually was not familiar with the procedures in the department being audited. They had no biases or could not make any assumptions about the procedures they audited; they had to be convinced during the audit by a showing of evidence that the procedures were being followed. Another benefit was that the process of having to explain what you do to someone who does not necessarily understand your procedure, but is challenging it, finds the errors or gaps that allow for improvements to be made. It's amazing what an unfamiliar pair of eyes can find! This is really the value of the internal audit process. Its continuous improvement being driven by continuous review of the system. When every finding and nonconformance from an internal audit is logged into the corrective/preventive action process, the system continues to grow and improve.
—Victor V. Margiotta, Director of Quality, SOBE

Achieving certification was positively worth the effort! The certification as an entity was not the benefit, the plus was that the internal quality audits mandated

by the system forced people into a number of good practices, which they may have otherwise let ship.
—Dave Demone, (Environmental and Quality Control Manager) & Sylvia Garcia, (Environmental and Quality Control Manager), Domino Sugar

The following are critical to an effective audit process:

- Top management commitment to the entire program.
- Management's role and responsibility clearly defined.
- Procedures that clearly define all requirements for performance of the audits.
- A responsible individual assigned to manage the complete program including scheduling, performing, and monitoring the audit function.
- Adequate resources available and applied to effectively perform all audit activities. This includes providing sufficient time for auditors to complete all their audit responsibilities.
- Audits planned/scheduled that include each aspect of the quality management program and at defined frequencies. Depending on the importance and potential risk sufficient to "verify whether quality activity and related results comply with planned arrangements and to confirm the effectiveness of the quality system." [ISO 9001:1994, Section 4.17]
- An audit plan that defines the scope and provides guidance to the audit team for:
  - Effective performance of the audit.
  - Verification and validation of the audit program.
- On-time performance of audits.
- Audits performed by individuals not having direct responsibility for the area or function being audited.
- Auditor training in ISO requirements and auditor techniques.
- Commitment from responsible management to address findings in a timely manner.
- Corrective actions tracked for timeliness and evaluated through follow-up activities to confirm effectiveness of the actions taken.
- Quality records identified and maintained to demonstrate compliance to defined requirements.
- Audit activities and results evaluated at the management review meetings with required actions implemented and resources applied, as appropriate.

Top Management Commitment

The audit function must be supported from top management through all levels of the operation. Management must clearly communicate that the audit function is a positive activity and not a fault-finding exercise. Management should communicate this commitment personally to all associates. Associates must be encouraged to address audit findings as areas of opportunities and potential "weaknesses" that, when addressed, strengthen the system. Audit activities and findings must not be used as a means to assess blame.

The audit program must have adequate resources. "Resources" may be defined as sufficient numbers of trained auditors provided with the time needed for the performance and maintenance of an effective audit program. The resource of time and availability must also be provided to the auditee. This must not be just a paper promise. The audit manager, audit team, and the auditee must be provided adequate time and support for the effective performance of all audit activities. An accurate accountability and status of the audit program must be reported to executive management at the management review meetings. This review should be conducted in a positive manner defining resource allocation and other appropriate assistance to ensure the effectiveness of audit activities.

> In an effort to display quality system commitment to all employees, the entire management staff of our company have become internal auditors. Even the president of the company is participating as an internal auditor. The results have been substantial. By completing internal audits, the management group has forfeited the relative isolation of the offices and are out on the "front lines" to see exactly what is going on. When inefficiencies are discovered the managers are in a position to accommodate a quick resolution. The employees see the manager on the floor and they get an idea of the importance of the quality system at the top of the organization.
> —Russ Marchiando, Quality System Manager, Wixon Fontarome

Managing the Audit Process

Management should assign a competent person as the audit manager or coordinator with the authority and responsibility to manage audit activities. This may be included in the management representative's responsibility by providing that position with direct involvement with audit performance. An accurate accountability and status of the audit program must be reported to executive management at the management review meetings. The discussion of the audit program at these meetings should include at a minimum a summary of on-time performance, any trends in the findings, resource issues both auditor and auditee, and timeliness of the corrective actions taken.

The Audit Schedule

Requirements for the creation and maintenance of the audit schedule must be defined. The schedule must include all aspects of the quality management program and ensure that each element of the ISO standard is audited at a defined frequency based on the importance of and risk to the area. Some programs split the system into individual audit sections; others do the complete system each time. Which is chosen will depend on the system's size, resources, and scope.

> It is important to realize that, like the Quality System as a whole, the Internal Audit process must change as the quality system matures. The primary focus of the audits to a newly certified organization will likely be on compliance to procedures and work instructions. This will certainly help build the discipline necessary to maintain an ISO 9001 system. However, as the quality system matures, the likelihood that documentation is up-to-date will increase significantly to the point of having the internal audits focus more on how to improve the process rather than process or documentation compliance. In addition, the focus of the audits may move to a departmental focus from an ISO element focus. This type of audit focus is, to some extent, easier to maintain and it provides for less duplication of efforts.
> —Tim Sonntag, Vice President of Quality and Technical, Wixon Fontarome

Ideally, areas to be audited should include a combination of both ISO elements and individual processes. For example, multiple elements may apply to a process or manufacturing area, whereas an audit of the document control process may be focused more on that specific activity. Even though a specific audit may be scheduled to evaluate document control activities, an audit of a manufacturing area would also include evaluation of document control requirements as related to that area.

> The approach to performing audits from an element or process standpoint is a very good one. We began doing this approximately one year after we were certified. It was a fundamentally different auditing style from what our auditors were used to. It forces auditors to sharpen their skills and take a "big picture" approach to the audit process. They now are asked to manage several elements at a time and form their questions to address these rather than focus on one element of the system. In the long run they will become better auditors because of it.
> —Russ Marchiando, Quality System Manager, Wixon Fontarome

To summarize, in addition to being evaluated as applicable in specific areas, ISO system-specific audits should be scheduled to include at a minimum such process systems as management responsibility, the quality system, document control, calibration, control of nonconforming product, corrective–preventive action, quality records, internal quality audits, and training. This will provide the opportunity to evaluate the management of the specific defined requirements for those activities.

The schedule should be created and maintained in a manner that tracks each audit. This can be accomplished through a matrix that identifies each area to be audited along with the specific assigned time periods for performing the audit. This matrix provides an excellent record to demonstrate that each element has been evaluated.

It is recommended that audits of each area and each ISO element be performed at a minimum of once per 6-month period (semiannually). Effective systems do sustain on yearly audits; however, much can happen in a 12-month period. It is much more advantageous to the system as a whole to perform audits with smaller scopes either semiannually or quarterly than it is to perform one large audit annually. A large annual audit may become burdensome in time constraints for performance, in report writing, and in responding to audit findings. Also audits with a smaller scope can be extensions of each other. This provides an excellent opportunity for audit findings to contribute to continuous improvement.

The ISO standard does not require a specific frequency for the performance of the audits. However, the frequencies must be based on the importance and risk of the particular area with audit results being reviewed as part of the management review process. When identifying the time periods, be careful not to create a schedule with too many constraints. If a specific day is scheduled then the audit must be done on that day. It is much more practical to focus on a specific week or month. This builds flexibility into the schedule.

Scheduling audits by the month that they are to be performed provides the auditor and the auditee ample opportunity to identify a time frame convenient to both groups. If an audit is unable to be performed during the scheduled month, then the responsible auditor should notify, in writing, the audit manager, stating the reason for the delay. At the discretion of the audit manager based on the importance and risk of the particular area to be audited, the audit will either be rescheduled or assigned to a different audit team. These requirements must be defined in related procedures or work instructions.

Related procedures or work instructions should also define the acceptable time frame for the performance of the audits to constitute "on-time". This may be during the month scheduled or plus or minus one week of the assigned date. If there is any deviation from this time period then this must be approved by the audit manager.

Performing "on-time" audits must be taken seriously with every effort made to perform the audits as scheduled. The audit schedule may be maintained as a living document coded to identify performed, in-process, completed, and rescheduled audits. The code may also be used to identify the audit team. Requirements for rescheduling audits should be defined and identified as such on the schedule. It is essential to the effectiveness of the system to be able to track audit performance. Develop a realistic schedule that does not burden the system but contributes to the overall usefulness of the system.

Some systems have found it effective not to continually update the schedule itself finding that having to revise it as audits are performed makes controlling it a burdensome exercise. Instead an internal audit performance log is maintained by the audit coordinator. This log tracks performance, timeliness, and any problems that may be encountered. Depending on how the process is defined, it may also be used to track findings and observations identified during the audits.

Remember the process is yours to define. Be careful not to box the process in by how it is defined. Build in some flexibility. Audit requirements must be clearly defined. However, as the system matures, revisions to the process as deemed effective and useful for the system as a whole must be implemented.

To provide adequate proof that the audits are being planned, the audit schedule should extend over a 12-month period. The thought patterns (the importance of and risk to the particular area) for determining frequencies for specific areas may change through the course of a 12-month period. For example, in one system, the retired purchasing manager was replaced with an associate from outside the company. Although an audit of the purchasing department was not scheduled for 4 months, the audit manager made the decision to add an audit of the area 2 months after the change to provide an opportunity to confirm that purchasing requirements had been effectively maintained through this transition.

The auditors must be independent of the area being audited. "Independent" is defined as not having direct responsibility for the areas being audited. This is important in that it provides each area an opportunity to be evaluated by a "fresh set of eyes." Many times those responsible or working in the area are too close to the activities and truly "can't see the forest for the trees."

Defining the Audit Scope

The scope for each audit should be clearly defined and communicated between the audit activity manager, the audit team, and the auditee (the group to be audited). This basically defines the starting point, what is to be evaluated, and the projected end point for the audit. For example, an audit of a blending area would identify the blending area, area procedures, work instructions, and related ISO elements such as training, calibration, quality records, and document control. When creating the scope, review historical information relating to the area to be audited such as previous audit reports, corrective/preventive actions, customer complaints, and the like. These will provide information on activities that may need additional follow-up or that may not have been recently evaluated in some time. This relationship between audits contributes to establishing the framework for a strong and effective long-term program.

Audit Planning and Preparation

Audit planning is also referred to as audit preparation. Effective preparation is important to the overall success of the audit activity. The audit team should review all documentation related to the scope of the audit. Examples of this would be previous audit results, nonconformances, corrective actions, industry regulations, technical data, and customer requirements.

INTERNAL QUALITY AUDITS

Preparation activities should result in the creation of the audit checklist. This checklist is an important tool to the audit function. This promotes a thorough, effective, and uniform audit. Generic checklists are available for almost every aspect of any audit. These are good for reference purposes; however, the audit team should prepare a checklist that focuses on the specific scope of their specific audit assignment.

The Audit Checklist

The audit checklist has many purposes:

- Clearly defines the objectives.
- Provides guidance for keeping the audit on track through preplanned questions and review of appropriate documentation.
- Aids in defining the sample.
- Becomes an important part of the audit record.

To prepare a checklist, the following should be done:

- Review the requirements of the ISO standard.
  - Have the standard with you for reference during the audit.
- Review related quality manual sections that apply to the scope of the audit.
  - Quality policy statement and the like.
- Review related procedures, work instructions, and other sources of information, such as:
  - Previous audits.
  - Regulations.
  - Known quality problems.
  - Outstanding corrective and preventive actions.
  - Management priorities and customer requirements.
  - Company information (brochures, new letters, etc.).
  - Auditor's background and experiences.
- Have the related documents readily accessible.

Keep in mind the checklist may be merely a handwritten list of questions. Where allowed by your system, the auditors may want an uncontrolled version of the appropriate procedures and work instructions for their reference. These must be clearly identified as uncontrolled documents. Review requirements for uncontrolled documents as discussed for document and data control.

Russ Marchiando (Quality System Manager, Wixon Fontarome) offered some interesting advice on auditing that he learned from a third-party auditor.

(Note that the third-party auditor is generally the term used to refer to the auditor representing the registrar.)

> Our business is very complex, with many different things going on at any time. As a result our quality system contains many different documents outlining each process. Sometimes our auditors feel a little overwhelmed by reading through a considerable stack of applicable documentation. I've learned something from watching our third party auditor that I've passed on to our auditors. That is that you don't necessarily need to read all of the documentation that is applicable to the situation. You just need to have an idea of what questions you want to ask and let the auditee provide you with the answers and the applicable documentation that the answers are contained within. I'm not saying that you don't need to prepare, but I am saying that the onus should be put on the auditee. They after all should know what is contained in their quality system documents. Finding this middle road takes a little of the pressure off of our auditors.

Performing the Audit

Actual performance of the audit brings together all the key points previously discussed in this text such as:

- Management communicating their support to associates at all levels of the operation.
- Trained auditors independent of the function to be audited performing scheduled audits.
- Efficient audit planning resulting in effective performance of the audit.
- A positive audit focus that is not an effort to find things wrong and assess blame but to verify and validate that defined activities are in place and effective.

Prior to actually starting the audit, the audit team should make arrangements to meet with the responsible management of the area to be audited. This provides the opportunity to review the scope and any other pertinent information. Examples of pertinent information would be associate availability, identification of the person(s) to accompany the audit team, and agreement of a time for the closing meeting to discuss the audit findings.

A responsible representative from the area being audited should accompany each auditor for the full duration of the audit. An audit should never be performed without an area representor accompanying the auditor. It is very important that this person see what the auditor sees. This person can provide information vital to the overall audit scope. Do not leave anything to interpretation. Remember that the audit process is only as effective as the relationship between the auditor and auditee.

INTERNAL QUALITY AUDITS

Sampling of Some General Audit questions

The following questions apply to almost all general audit situations:

- Are there procedures and work instructions present to comply with the ISO 9001 standard and the quality system?
- Are the procedures and work instructions under document control?
- Do the procedures and work instructions reflect what is actually occurring?
- Are the associates who are responsible for carrying out these procedures familiar with the procedures and where they are located?
- Are the associates familiar with the quality policy, customer requirements, objectives, and goals of the facility and his or her role in adhering to them?
- Is there objective evidence (records demonstrating compliance) that proves that what is stated in the procedures or work instructions has actually occurred?
- Do identified nonconformances follow the related procedures?
- Are records identified and maintained according to defined requirements for quality records. (Review ISO 9001: 1994 Section 4.16)?
- Are the procedures effective and maintained under document control?
- Are there any external documents used or referenced?
  - Are they defined and maintained under document control?
- Is there any posted information or other documents referenced that should be in the controlled document system?

Phrases to remember:

- Where is that defined or written?
- Please show me.

The auditor should develop a degree of rapport with the auditee but remain objective and remember to collect all essential facts. When the facts indicate a nonconforming situation, the findings should be reported in a tactful but meaningful manner. This is absolutely not a fault-finding exercise. Facts should be listed clearly. Conclusions should be based on facts. Always remember that the audit is a sample, and sampling activities do have limitations. The auditors should keep a positive attitude, creating an atmosphere that promotes good communication. Always thank the auditees for their assistance and time regardless of the outcome. Auditees should never feel threatened by the auditor. An auditor should always treat the auditee as he or she would want to be treated.

Audit activities should focus on the scope, with the auditors making clear audit notes as to the questions asked, the location of the related materials, any observations and concerns. However, many times unplanned trails that must be investigated surface during an audit. It will depend on the situation and at the discretion of the audit team whether time is spent investigating during the current audit or during a subsequent audit. For example, the auditors may identify a potential training issue in the receiving area. Depending on the situation they may follow that trail during this audit and/or provide the necessary information to have it or a similar situation investigated in a subsequent audit.

These situations are of course unforeseen in the planning phase, but should be given appropriate attention. The audit report should note such findings and the ultimate decision. By noting this, the record provides subsequent auditors the opportunity to follow-up on areas originally planned, but preempted in order to address unplanned activities essential to the audit activity.

Auditee comments, opinions, and hearsay information may be noted informally and investigated, but only included in the actual report if the auditor can substantiate it by fact. Do not mention names or include he said/she said comments. Edward Link in *An ISO 9000 Pocket Guide for Every Employee* states that

> "The Quality System must always remain the focus of the audit. Regardless of [the] answers, the auditor never sees the individual as the root cause. A good auditor will never approve a corrective action that in any way suggests that the individual was at fault." (Page 83)

Thoughts on "audit notes"

- Record all relevant facts seen and heard.
- Be patient as you develop the technique.
- Record sufficient facts to make an informed judgment.
- Clearly identify/reference pertinent documents, batch numbers, order numbers, etc.
- Notes should be legible and understandable.
- Identify relevant areas/trails, etc.

There should never be any surprises at the closing meeting or in the audit report. Auditor's concerns should be discussed on the spot in a positive manner with the auditees and the area guide. Many times answers can be provided; other times the situation may not be as initially observed. The guide or area representative can provide the linkage to the information required.

Grading Noncompliances

Some processes define that auditors must grade the findings. Examples would be "major," "minor," and "observations." This works in some situations, but

INTERNAL QUALITY AUDITS

ensure that procedures and work instructions have a clear definition of these grades and that auditors continually review these. This type grading system can become very subjective, causing confusion within the system. Most systems, as they mature, realize that a more effective means is to not grade the nonconforming situations. The "time-frame" commitment for completion would reflect the priority of the finding. Observations should also be investigated.

A nonconformance is defined as a situation that violates a requirement of the standard or a defined requirement of the system. Examples of frequent nonconformances are included at the end of this chapter. An observation may be defined as either a situation that if not addressed would likely result in a nonconformance or a suggestion to improve the process or system. An observation may also be reported when the auditor suspects that there could be a nonconformance but, at the time, could not generate the proof to report it as such. For example: "Although audits are now being performed on time, it appeared that the decrease in number of auditors could result in difficulty scheduling and adhering to on-time performance over an extended time frame."

It is very important to the strength and effectiveness of the system that the responsible area managers investigate not only the nonconformances but also observations. Remember an observation may be a warning of a potential problem. Nonconformances and resulting corrective actions are discussed in more detail in Section 9.4 of this chapter.

Thoughts to remember about the audit:

- The audit is a sampling exercise.
- It is not meant to find everything that is wrong.
- It is a snapshot in time.
- Findings are based on factual evidence.
- Focus in on the system.
- It is not a faultfinding exercise.

The closing meeting and the final audit report

The closing meeting should be held upon completion of the audit. The purpose of this meeting is to discuss the findings with the responsible area management and as many area representatives as possible. Some processes require that the audit team prepare an extensive report. Every effort should be made to prepare this report as the audit is being performed. Ideally, it is best to submit the written findings at the closing meeting; however, depending on the particular situation, this may not be possible. If this is not possible the written report should be completed as soon as possible after the completion of the audit and no more than one week from the closing meeting. Any longer than this, decreases focus and effectiveness of the audit.

Remember that audit findings are areas of opportunity to strengthen the system. Findings must never focus on assessing blame. Do not mention names in the report. Findings must be based on fact, never include hearsay information. Findings should be reported and documented constructively in a predetermined format. Responsible area management should ensure that root cause analysis and corrective actions are taken in a timely manner.

Surviving the Audit

As the auditee, there will be many challenges. Always be prepared, take your time, and if you don't know the answer to a question just say so, don't make up answers or try to give the auditor what you think he or she wants to hear. Be honest and don't try to adjust the answer. Experienced auditors will detect dishonesty. Be careful not to answer questions that are not asked; however, be helpful. Be confident, after all no one knows your job better than you do. Do not argue with the auditor. There is a difference between arguing and challenging. The auditee should challenge the auditor if this challenge is based on fact. Do not argue just to argue or waste time. Use the following as guidance:

Always Be Prepared

- Know your job and related customer requirements, and understand your responsibilities.
- Know related procedures and work instructions.
- Know where they are kept in your area.
- Know what records are within your responsibility.
- Know the retention times and retention locations for those records within your responsibility.
- Know how you were trained.
- Know the quality policy and how it relates to your responsibility. ("Doing the best you possibly can.")

Take Your Time

- Don't rush to answer.
- Be sure you understand the question.
- If you don't understand the question say so and ask to have it repeated.
- Get clarification, if necessary.
- Always think about the answer before responding. Be honest.

After the Audit

The audit program should include a follow-up evaluation to ensure that corrective actions taken were effective. Many programs actually schedule specific audits at defined frequencies such as monthly or quarterly to provide the opportunity for trained auditors to complete this evaluation. Be sure that enough time and evidence is available to confirm effectiveness. For example, if the root cause is identified as insufficient training and the corrective action is to train, then significant evidence must be available to confirm that the training was effective. Confirmation of effectiveness is not merely the completion of the training exercise, but a review of the activity, records, and any other appropriate information necessary to confirm that the activity is now in compliance.

Some systems define that audit corrective actions are incorporated and tracked through the systemwide corrective/preventive action process. In this particular situation the audit may be considered closed once the report is issued. Although a system can successfully manage the process with the audit corrective actions handled separately, generally over time most find that it is best to have one corrective/preventive action process. Responsible associates find it more effective to manage, track, and monitor through a centralized system. This must be a management decision. It may actually be best to handle them separately in the beginning, combining the activities as the system matures. The corrective/preventive action process was discussed in more detail in Section 9.4.

Training the Audit Team

The training that the internal auditors receive will very likely be directly proportioned to the overall effectiveness of the audit program. Training criteria should include the attendance to an accredited audit course. Auditors must be trained not only in ISO 9001 requirements (current version) but more importantly in auditor techniques and protocol. Auditors must be trained in the positive attributes of being an auditor. The auditor must understand that his or her role is to act as an extra set of eyes to confirm and reinforce the effectiveness of the total quality management system.

Two-day internal auditor training courses are very thorough, teaching the use of the ISO standard and important auditing techniques. Attending an external course is beneficial, however, there can be even more benefits realized by having the course taught on-site. Instructors can adjust the course material and perform structured "teaching" audits of your organization as part of the course's activities. An external course can be confusing to food industry professionals because many times the course material will focus on totally unrelated industries. Well-planned on-site classes can be more economical and increase the opportunity for attendance.

The purpose of auditor training is to provide a strong foundation for understanding requirements of the ISO-compliant system. This level of exposure may include attendance to a comprehensive 5-day lead auditor course by at least two team leaders such as the management representative and his or her alternate. An on-site internal auditor-training course for the remainder of the audit team may be developed from this learning experience. A copy of each auditor's certificate confirming attendance should be maintained as a quality record.

Related audit procedures and/or work instructions must define the required training for the internal auditors and the quality records that must be maintained to demonstrate compliance.

More on Auditor Techniques

Auditors must focus on a positive performance style. Auditors should always treat auditees as they would want to be treated. Nothing causes the program to fail quicker than to have an auditor that presents him or herself as an overpowering force. People skills of the auditor are as important if not more important than knowledge of the ISO standard. One doesn't have to memorize a dictionary to know how to use it. Using the ISO standard is the same philosophy. Even professional ISO auditors that use the standard every day do not specifically memorize it but know how to reference it.

Auditor Requirements

An auditor should be

- A good communicator, a good listener, and a good organizer, and exhibit a professional image.
- Patient, tolerant, polite, and confident.
- Objective, persevering, and have a positive attitude.
- Precise, but practical, and practice interpersonal skills.
- Nonthreatening.
- Non-judgmental and opinionated.
- Empathetic (not sympathetic).
- Able to validate "tip-offs."
- Open avoiding secrecy.

Stress that auditors approach the auditee in a positive respectable manner. After all, no one knows the process better than that person performing the activity (i.e., the operator). An auditor's focus is to confirm compliance. The success of the audit will be greatly hampered if the auditor enters the area to be audited as a "bull in a china shop." The auditor must understand that his or her role is to act as an extra set of independent eyes to confirm and reinforce the effectiveness of the total program. It is not to find as many things wrong as possible or to get people in trouble. Never lose sight that the audit is a fact-finding not a fault-finding exercise. The auditor should always treat the auditee with the same respect that he or she would want to be treated.

Audit Team

The audit team should consist of a diverse group of associates representing all levels of the operation. Ask for volunteers from the work force. Being part of the audit team is an excellent opportunity for everyone to not only learn about other aspects of the operation but to contribute ideas and motivation to the system. Individuals within the organization tend to share enthusiastically the auditing experience with their peers. In addition, ideas and knowledge may be brought back to their own departments to be applied and to strengthen its activities.

Once the team has been identified and trained, plan periodic team meetings. Use these meetings as an opportunity to discuss findings, to refresh techniques, and to receive feedback from the auditors. Many organizations plan quarterly pizza lunches for these team meetings.

Remember that resource availability for both the auditors and auditees is a must to an effective quality management system:

> We have had difficulty with auditors finding the opportunities in their schedules to allow time to audit. Often times when they do find time to audit, our auditors become a little overzealous and want to identify all of the company's shortcomings in one day. This inevitably results in a large amount of paperwork and a feeling of unpleasantness and distrust with the audit. This problem seems to be dissipating as the system matures.
> —Russ Marchiando, Quality System Manager, Wixon Fontarome

An opportunity may arise to have the system audited by an external auditor to your system. This may be an associate from a sister plant or by a consultant hired to evaluate the system for compliance and improvement opportunities. These audits can be invaluable to the total audit program by presenting a different viewpoint from a different source of independency. In order to include these audits in the internal audit program, define in the appropriate procedure or work instruction that these types of audits are acceptable as long as an auditor performs them that meets the defined auditor training criteria. These audits should be identified on the audit schedule. The auditor's credentials demonstrating compliance with defined training requirements must be maintained as quality records.

Audit Records

The internal audit record should include audit notes, the checklist, a summary of the findings, and the report including observations and noncompliances. This information will be invaluable to the planning phase of subsequent audits.

Auditor training records may be maintained as part of the internal audit function or with the other system training records. Internal audit procedures or work instructions must identify which records are required and where these are defined (i.e., reference the systemwide quality record procedure).

Some Benefits Experienced from an Effective Audit Program

- An increase in awareness of quality among personnel.
- Enhanced quality system development and continuous improvement opportunities.
- The identification of improvement opportunities.
- Objective feedback based on facts.
- Information for effective allocation of resources.
- Early detection of potential problem areas.
- Meeting the requirements of the ISO 9001 standard for internal quality audits.
- Confirmation that customer requirements are being met.

Remember, the benefits from an extra pair of eyes independent of the area being audited is one of the most valuable sources of information available to your quality management system.

More Thoughts on the Internal Quality Audits

> We took an unusual approach at internal audits and started them very early in the quality system implementation effort. By starting the audits so early in the process we felt that we could take advantage of a good opportunity to identify potential implementation weaknesses as well as allow our auditors to gain confidence and become seasoned by the time we were ready for certification.
> —Russ Marchiando, Quality System Manager, Wixon Fontarome

> Our auditing process has probably been the most useful because it has broken down so many barriers and has led to enormous improvements throughout the organization. This has occurred in ways it was not even designed to do. Besides the obvious benefits of auditing, our process has led to a large amount of [employee] cross-training, team-based action plans, and an increase in intra-departmental communication and consistency. Our auditing program over the past three years has progressed to auditing for effectiveness and efficiency, rather than auditing to meet requirements.
> —Keith Gasser, Quality Systems Manager, Tropicana

> After certification, one of the toughest aspects of the system, is to keep the auditing function challenging and growing as the system matures. Although the quality of the audits continues to improve, it seems like the auditors find it more and more difficult to find the time to perform the audits as scheduled. Prioritizing and communicating the importance of timely audits and timely response to audit findings is an important management review opportunity.
> —Karen Morgart, Packaging Specialist, Hiram Walker & Sons, Ltd.

> Quality auditing provides a disciplined, proactive approach to system improvement. Audits are an excellent tool for identifying weaknesses before they manifest themselves as costly failures.
> —Eric Halvorsen, Quality Assurance Manager—Auditing, Campbell Soup Co.

In Conclusion

This section provides a generic overview for establishing and maintaining the audit function of an effective quality management system and can be applied to many different processes. Each system should develop its own program specifically designed for its needs backed with top management support focused on team involvement. A baseball pitcher could not pitch a no-hitter without eight players on the field and a full roster of trained and competent players on the bench ready to do what they do best, the best they can.

Edward Link (1997) in *An ISO 9000 Pocket Guide for Every Employee* sums up the thoughts on internal auditing very well in the following quotation:

> Management's Role—The investment made in the quality system will provide significant pay back if the system is kept in a healthy state. The internal audit subsystem is the self-check necessary to maintain quality system health. The effort in conjunction with the third party audit conducted by your registrar virtually assures a highly effective quality system in the short term and the long term. Recognizing the benefit and supporting the effort with the necessary resources is required of management.... Each Employee's Role—In this area, everyone is involved as an auditee and/or an auditor. As an auditee, be prepared to be totally open with your internal auditors so that the measurement being made on the health of the quality system is as accurate as can be despite its often subjective nature. Internal auditors should be mindful of the training that was provided to you as you prepared to become an internal auditor. If you are unsure about the exact interpretation of a requirement, seek the help of more knowledgeable and experienced users of the standard. Declare yourself ineligible to audit any area where you think that there is possibility that you could not be both fair and thorough. Do everything that you can to make the measurement you are making as accurate as possible. (pp. 69–70)

Frequently Identified Nonconformances

- Audits are not being performed as scheduled.
- The majority of audits performed are late, with or without an explanation as to why they are late.
- Training criteria for auditors is either not defined or records are not available to confirm that auditors have met the defined criteria.
- Records of auditor training do not provide evidence of training in the current version of the ISO standard for which the system is approved and operating.
- The audit schedule is either not being maintained or maintained in a manner that does not show evidence that audits are planned.
- The required frequency for performing audits is not defined.
- Training criteria for internal auditors does not identify training on the ISO 9001 standard as a requirement.

- There is no evidence that all the elements and processes are being audited within the defined time frame.
- Review of the corrective actions issued from the audits did not provide evidence of one or more of the following:
  - Reported to responsible management and/or addressed in a timely manner.
  - Completed corrective actions are evaluated for the effectiveness of the actions taken.
- Audit record requirements are defined; but records are not being maintained as required.

9.6 STATISTICAL TECHNIQUES

The standard requires that where statistical techniques are used to control the process requirements must be defined as related to the organization's activities. Defined requirements should include the responsibilities for performing the activities and for handling analysis of the results. Examples of areas where statistical methods also referred to as statistical process control (SPC) may be in evaluating process capabilities, receiving inspections, on-line sampling, and quality control analysis.

The requirement and extent that statistical techniques are used in different industries vary considerably. Robert Peach (1997) in *ISO 9000 Handbook, Third Edition*, states:

> Each company must determine what type of statistical techniques, if any, are appropriate based on customer requirements, industry practice, and overall cost in relation to the product. (p. 161)

For the most part, in the food industry, companies that use statistics use them more for monitoring the process and as an improvement tool rather than for the "go–no-go" decision. Unlike some industries that require statistical techniques to identify upper and lower acceptance criteria, food manufacturing processes confirm almost all specification compliance through some form of testing.

Edward Link (1997) in *An ISO 9000 Pocket Guide for Every Employee* states that in discussing the quality system, quality has been thought of in terms of

> reduction of variation. The most often used quote of Dr. W. Edwards Deming is "There is no substitute for knowledge." His teachings have helped us realize that we must first understand variation before we can reduce it. Statistical Techniques enables the understanding of variation. This element is unique in that it directs the [organization] to first conduct an assessment of need for statistical tech-

niques. . . . when utilized, statistical techniques must be implemented and controlled using documented procedures. (pp. 77–79)

If statistical techniques are applied, then in these instances it is common to define the requirements as part of process control or inspection and testing procedures. The quality manual should provide the reference (road map) to where the appropriate procedures or work instructions for these requirements are actually defined.

Statistical techniques performed for process monitoring and process improvement opportunities must be clearly defined as such. Do not get overzealous and define stringent requirements that are not practical and in fact not being performed. An auditor will evaluate the process including the thought pattern and justification for the decisions made in addressing the requirements for statistical techniques.

It is common in the food industry for a system that doesn't use statistical techniques to simply state in its quality manual that "at this time statistical techniques are not being used and are not required within the operation." Do not state that this does not apply, but rather at this time the system does not perform these types of activities. In addition, a statement should be made that includes a periodic review of the status such as annually at a management review. This should be included on the management review agenda with results of the discussion recorded in the minutes (quality record).

Frequently Identified Nonconformances
- Procedures have not been established for statistical techniques used to control the process. (When the system was first implemented statistical techniques were not in use, and the quality manual and related documents defined this as such. However, as the system matured its use was implemented. Documentation was never revised to reflect this.)
- It was stated that statistical techniques are used to control the processing operation in regard to weight control of the finished product; however, records are not available to demonstrate that readings are actually within the upper and lower control limits.
- Review of a sampling of records indicated that actual readings are frequently outside of the upper and lower control limits with no notations as to what had been done to address these situations.

10
CERTIFICATION

The certification effort may seem at times like being on a long dark road wondering where in the world is the destination that you have worked so hard to achieve. Once a company has achieved certification, then its efforts must focus on maintaining the system and achieving continuous improvements through the system's maturation and growth. ISO compliance does become a way of life. Associates and management tell me frequently that they don't know how they managed without it.

What Has Compliance Meant to Your Process?

Compliance has meant so much to so many. Following is a wide variety of comments that basically have "consistency," "accountability," "commitment," and "improvement opportunities" in common.

> ISO compliance has instilled a level of accountability in thoughts, processes, and actions throughout the company. ISO has provided a reference point for planning all new products and processes.
> —Jim Murphy, Manager of Design Process and Validation, The Dannon Company

> Compliance within our system has meant consistency and discipline internally; credibility and quality externally. It has enhanced our ability to managing change. Business strategies, organization structures, products, processes, systems, personnel, and philosophies all change very rapidly. Interestingly, now our ISO 9001 based quality system is much more able to adapt and control these changes than

ever before. We have the structured discipline to adjust our systems in conjunction with whatever change occurs.
—Keith Gasser, Quality Systems Manager, Tropicana

Summing up in one or two phrases what compliance has meant to our organization would be that it gave us the ability to restructure the organization, starting over with only administrative staff and being able to manufacture products from the documented procedures, work instructions and formulations.
—Yvette Castell, Quality Assurance Manager, Dairy Industries (Jamaica) Limited

With our quality management system in place, we have been able to tackle new challenges head on, and use the process to effectively implement the required standards. It has stood firm through re-branding, restructuring, change in General Manager and a most challenging business environment.
—Linda Taylor, Training and Quality Manager, Le Meridien Jamaica Pegasus

Compliance to the requirements of the standard has meant a higher level of executional consistency in our manufacturing facilities. This has resulted in higher product quality while maintaining consistent employee performance, consistent system development, and most importantly, a disciplined and documented method for improvement (management review, corrective and preventive action, etc.).
—Mike Burness, Director of Quality Assurance, Pepperidge Farm Inc.

Certification has provided our quality system with structure, definition and a means of resolution on our path towards continuous improvement.
—Russ Marchiando, Quality Systems Coordinator, Wixon Fontarome

We now have a structured and disciplined system for delivering services that consistently meet management and internal requirements. The system is flexible enough to easily adapt to changes in management and customer directions. It also gives us a good foundation or platform for implementing processes that are compliant with business excellence models such as the Baldridge Criteria.
—Tom Marchisello, Director of Quality Assurance, Campbell Soup Company

Consistency and discipline are the words that most often apply.
—Rick Aldi, Director, Quality & Environmental Affairs, Hiram Walker & Sons, Ltd.

Compliance to us has meant a couple of different things, one, is that we are kept within quality boundaries and the second is having reliable records to demonstrate due diligence!
—Henry Gibson, Quality Assurance Manager, Campbell Soup Company

Bill Lockwood (Package Quality Manager, Hiram Walker & Sons, Ltd.) when asked what ISO meant to his operation stated:

ISO gives us the initiative to follow good business practices. [It can be said that] ISO is the backbone of our quality management system, the [elements] are the body, and the external auditor is the key to commitment. ISO has made a tremendous impact on [our company]. When we were finishing our Leadership

Training, it pulled everyone together to reach a common goal, giving us the opportunity to use our newly learned leadership skills.

Compliance Also Improved Customer Relations

One of the greatest most useful benefits that we achieved from certification was the development of a system supported by procedures that permitted customer audits to flow smoothly providing a sense of assurance that we were in control and understood both our business and their needs.
—Dave Demone, Environmental and Quality Control Manager, and Sylvia Garcia, Environmental and Quality Control Manager, Domino Sugar

Andy Fowler (Research Engineer/Management Representative, Bacardi & Company Limited) stated that compliance has resulted in a system that operates "a lot smoother and a lot more efficiently." Gail Cartwright (Assistant to the AVP Human Resources Department, Bacardi & Company Limited) stated that she feels compliance makes "the whole process a lot more structured allowing implementation of new ideas and technology."

More Thoughts on Compliance Promoting Consistency Within the Process

Compliance has helped us do things the same way and to question ourselves and others when we stray away from the procedures.
—Charlie Stecher, Quality Assurance Manager, Reckitt & Colman, Inc.

If I could sum up what compliance has meant to our process it would be ISO 9001 has given us a structure to ensure consistency, improve processes and determine and measure our goals.
—Sue Goode, ISO Coordinator, Cargill Corn Milling

The ISO 9001 certification process and continued compliance to the standards has provided us with not only a well documented, structured quality system with well-defined responsibilities, but it is apparent that it can mature into a "way of doing business" that can drive continuous improvement throughout the organization.
—Tim Sonntag, VP Quality Assurance & Technical, Wixon Fontarome

System-Enhanced Training Activities and Promoted Teamwork Throughout Organization

The best thing about certification and our quality system was that it defined the ability to train operators on a consistent basis and slowly develop the concept that only in-specification product was shipped every time. No cutting of corners, no back room decisions or short cuts. Compliance made us identify trouble spots, pay attention to them, and take actions before they became critical. It also promoted a true sense of teamwork and a common purpose, particularly in the earlier days of seeking and obtaining registration. Most employees ultimately

understood and supported the system as it provided a degree of stability to their job. The system "worked" for them, not against them!
—Dave Demone, Environmental and Quality Control Manager, Domino Sugar

ISO by its own nature needs continuous improvement to ensure that it continues to improve. It must sustain on what it itself preaches. We continue to improve through training, continuous improvement, consistent procedures, [and] follow-up.
—Bill Lockwood, Package Quality Manager, Hiram Walker & Sons, Ltd.

Most Difficult Aspects of Maintaining the System Since Certification

Some participants stated that convincing middle management that maintaining ISO certification was as much of a team effort as achieving it was the most difficult aspect of the process. If the system is managed (maintained) by one department, such as quality assurance, then it is thought of as a quality assurance program. Many times this relates back to allowing middle management to dissociate itself as much as possible during the certification process, which expands into further detachment from the quality system after certification. This is a situation that must be addressed through management review and other system maintenance type of functions. It relates back to understanding that it is just as important to maintain team involvement after certification as it is during the implementation process. My favorite analogy is that of the baseball pitcher who cannot pitch a no-hitter by himself. It takes the full team made up of many individuals with many different skills all doing what they do best the best that they can. It is essential that associates from all levels take ownership for their role and associated accountability. Think about the outfielder who misjudges a fly ball. It could be said that he is not responsible for catching the fly ball. The pitcher should not have pitched a ball that could be hit into a fly ball. This is not the thought pattern for success.

One of the most difficult aspects of maintaining certification is to keep the focus and maintain the momentum. Following are some comments shared by professionals who have experienced the great feelings of accomplishment and the "growing pains" as the system "matures."

> The most difficult is always the same, keeping the interest of the people channeled and informed. From what I have seen there has been a shortage of good education on a continual basis to keep all persons informed and to keep interest in the system at a high level. One does not take graphs and numbers to a meeting to show to the employee. One must present a program that tells the employee how he or she will benefit. Informative meetings with drinks and snacks are received as a reward and therefore gain attention.
> —Jon Porter, President, J. Porter and Associates, Ltd.

> The most difficult aspect of maintaining the system after certification is communicating the message that the goal is not only to become certified but is to

continually improve the system. Most of our employees saw the finish line as the successful completion of the initial audit when realistically there is no actual finish line. Another difficult aspect of maintaining the system are the pains that are experienced as the system begins to mature and must be updated and refined. What may have been a necessity at the time of implementation may no longer serve the company well as the quality system matures. Everyone must understand that the quality system will always be in a state of flux as the company moves toward continuous improvement.

—Russ Marchiando, Quality Systems Coordinator, Wixon Fontarome

The most difficult aspect of system maintenance has been, without a doubt, keeping the purpose of the system in perspective. It is very easy to take a deep breath after certification and say that you are done. It is also deadly. The purpose of the system is to improve business effectiveness. Two cornerstones for attaining this are management review and corrective/preventive action. Maintaining focus on these items as they relate to the business, is the most difficult, and at the same time, most rewarding aspect of the system.

—Mike Burness, Director of Quality Assurance, Pepperidge Farm, Inc.

The most difficult part of maintaining the system is the initial sigh of relief and let down by almost everyone in the facility that the audit is over. It takes a strong level of management commitment to continue to show that the system is the way that the plant will be managed.

—Rick Bay, Plant Manager, Reckitt & Colman, Inc.

After certification there is a tendency to let down your guard and allow the system to slip back away from compliance. Our situation is not unique to this reaction and we certainly had a difficult time communicating the fact that our certification was just the beginning of the quality system's maturing process. As our Quality Systems Coordinator accurately stated on numerous occasions "ISO 9001 certification is a journey, not a destination." It is the beginning of a quality system that will be continuously changing and maturing. These changes could render some of the previously defined requirements and documentation as unnecessary as the system matures and moves towards company-wide continuous improvement. In addition we found that commitment and involvement from the top of our organization was needed and that the vehicle to maintain this level of commitment could not just be our quarterly Management Review Meetings. Communication needs to flow continuously from the ISO watchdogs (i.e. the QA Department) to the top of the organization to keep them abreast of the current status of the ISO system and to address issues that could threaten our certification. To assure this continual line of communication we have implemented a $\frac{1}{2}$ hour weekly meeting with the President of the company to discuss quality system issues. The results to date have been fantastic.

—Tim Sonntag, VP Quality Assurance & Technical, Wixon Fontarome

The most difficult aspect of maintaining certification has been in keeping up the momentum! After the initial hoopla upon certification, we became very complacent and almost collapsed the system.

—Rick Aldi, Director, Quality & Environmental Affairs,
Hiram Walker & Sons, Ltd.

CERTIFICATION

The most difficult aspect of maintaining the system after certification was getting people to regularly review their process documentation for currency, keeping their training records up-to-date and to spontaneously generate corrective/preventive actions.

—Yvette Castell, Quality Assurance Manager,
Dairy Industries (Jamaica) Limited

Tom Marchisello (Director of Quality Assurance, Campbell Soup Company) stated that one of their most difficult tasks since certification was "to keep management, internal customers and users of the system informed of the changes and their benefits." He also indicated that "it is difficult to keep up with training when constant changes are being made."

It has been difficult to maintain interest in the system after certification.
—Andy Fowler, Research Engineer/Management
Representative, Bacardi & Company Limited

A common concern was the complacency that may occur after certification. It is so easy for some to rely on the management representative or the managing department to maintain compliance. Complacency can and will collapse a system. This type of situation must be identified and resources assigned to bring it under control. Once the system is implemented, it is necessary to ensure that it continues to operate effectively. The system by design is maintained for suitability and effectiveness through the processes of corrective actions, preventive actions, internal quality audits, and management review. Keeping up the momentum can truly be the challenge, but in time through team effort and focus, system activities will become part of the everyday way of operating the business. It truly becomes a way of life.

Setting Goals and Measurable Objectives to Monitor Return on Investment

Return on investment can be defined in two primary ways. The first is based on a subjective comparison of where we were 3 years ago prior to our commitment to an ISO quality system and where we are now 18 months after certification. Subjectively, I feel we have obtained adequate return on investment for the age of our system. Since the number of ISO certified companies in the U.S. Food Industry is small but growing, we are in select company. I feel it has helped us to secure new and existing business, because from a quality system standpoint, the elements making up the ISO standards are exactly what many of the major food companies in the U.S. and abroad seek (along with effective food safety systems like HACCP, GMPs and thorough sanitation programs). Before our ISO commitment our quality system was much more departmentalized and was primarily a quality control department responsibility. The certification process helped us to implement a lot of programs that were desired by the Quality Control Department years earlier but were never implemented due to lack of organizational commitment. Quality is now being communicated throughout the organization and there is a good framework in place to bring forth company-wide improvement. From an objective return on investment basis, the jury is still out. In the last 12 months we have defined quality measurements to be taken and reported to upper levels of management on a quarterly basis. This will not only highlight areas of improvement needed but will also provide an objective measure for return on investment purposes. By tracking these measurements on a quarterly basis and comparing them to previous measurements we anticipate being able to track our level of improvement (which can be interpreted as an objective return on investment). In addition to utilizing these measurements in this way, the quarterly results and annual comparisons will be used to set organizational and/or departmental improvement-based goals. These goals will be tied directly to yearly goals and objective set for all managers and supervisors with the results discussed at semi-annual or annual performance reviews. As of now, there have been some goals established for some areas and departments in the company but others are still in the data gathering phase for establishing a baseline for future comparison. As a measurable objective, we have just started to identify and monitor the cost of producing unacceptable products that have to be either reworked or disposed of. This will then provide a quarterly cost of producing substandard quality product and its effect on the bottom line. The area responsible for the substandard product will also be tracked to set departmental or area goals.

—Tim Sonntag, VP Quality Assurance & Technical, Wixon Fontarome

"Achieving Certification Was . . . a Lot of Hard Work . . . , But an Invaluable Experience"

Achieving certification was the result of a lot of very hard work. In retrospect the experience was personally and professionally invaluable. The organization, has received a permanent benefit from the effort. The majority of employees at all levels see quality in a different light and context. I doubt that we'll ever ship out-of-specification product just because it is convenient or cost effective. The value of meeting the customer's needs is part of everyday life and that alone is worth the effort.

—Dave Demone, Environmental and Quality Control Manager, Domino Sugar

11
AUDITORS ARE HUMAN

In researching and preparing this text, many have asked me to include a chapter on what it is like to be a "third-party" auditor. For the purpose of this discussion, a third-party auditor is an auditor who performs ISO compliance audits on behalf of an accreditation registrar. Why do people become auditors? What do we do all day and how does it feel to be on the other side? What do we like best about auditing? What least? What is it that auditees do that we really like? What is it we really don't like? What do we do with all those frequent flyer miles?

There are, I am sure, many reasons why people become auditors. Probably if one asked 10 auditors, they would each have a different reason. My reason was my interest in the food science field. Auditing gave me a tremendous opportunity to learn about so many different processes. Opportunities in the world of third-party auditing depends on one's professional and "independent" quality system auditing experience. Independent means that one audits a system for which he or she has no direct or indirect responsibility. Unfortunately, it is tough for food industry professionals to get the independent status because so many of our audits are done for our employment company. If you are interested in being a third-party auditor, then it is recommended that you get as much ISO experience as possible through your company. You can also supplement your learning through professional organizations such as American Society for Quality (ASQ) and Institute of Food Technology (IFT). A successfully completed accredited ISO 9000 or 14000 (if that is your interest) five-day lead auditor training course is mandatory. Contact those registrars with food industry approvals (this information can be obtained from the McGraw-Hill Company), requesting information on opportunities with their companies. Although most registrars want the applicant to have achieved lead auditor status from either the International Register of Certified Auditors (IRCA) or the Registrar Accreditation Board (RAB), without the independent audits, this cannot be achieved. Many registrars will provide this opportunity as part of the training process. As the food industry becomes increasingly interested in ISO 9001 certification, the need for auditors with food processing experience is increasing. IRCA and RAB issue the lead auditor accreditation to an individual that provides the necessary credentials to confirm that competency. Each auditor is assigned specific industry codes that they are qualified to audit. These codes reflect one's education and work experience. The RAB is a private, not-for-profit organization and is an affiliate of the ASQ. RAB was established in 1989 to provide accreditation services for ISO 9000 Quality Management Systems. The RAB evaluates the credentials and assigns assessor or auditor grades as appropriate. Basically, a certified auditor may hold credentials from one or both of these organizations. The assessor must have these credentials in order to perform accreditation audits on behalf of a registrar.

Completing the registrar's training qualifications can be a humbling, frustrating, but very rewarding experience. As one auditor told me when I was going through the process, it is just something that you have to do so you might as well love it. Many registrars actually require that the majority of the training be completed in a field other than your field of expertise. This really teaches the trainee a strong understanding and application of the ISO standard. However, with a food science background, visiting foundries, telephone cable installers, and electronic manufacturers was an experience. After 3 months of training, I personally had seen enough ball-bearings and welding shops. I met some really nice people, found all the auditors that I had trained with terrific, gained knowledge and experience, but was ready to return to the food plants. Although I had successfully met all the training criteria, my training had really just begun. The primary focus of our assignments is to continue to learn, work with the auditees, and just do the very best we can to evaluate and report in a positive and constructive manner the compliance status of the organization being audited.

Having spent many years as a food industry professional, I knew what it was like to be audited. Thus I had a good idea of the type of auditor I did not want to become. Also having worked as a management representative for a food processing company that was going through ISO compliance, I had the greatest privilege to work closely and to actually become friends with our registrar's auditors. Watching them in action was a great inspiration to me. I saw firsthand the type of auditor that I knew that I would want to become. Fortunately for me, as the years went by, I had the opportunity to perform audits along side one of these auditors. Developing auditing skills takes practice. Auditing is an art that one never quits learning. Every day brings new meaning and new knowledge that never ceases. The ISO standard is very generic and as an auditor, one is trained to ensure that the auditee's system has been defined, is controlled, and being maintained in a compliant manner. Each system is different. Systems mature and change in different ways. We learn quickly that we have to stay focused and concentrate on the system being audited. There are still days when I feel that I concentrate so hard that my brain hurts. But the entire auditee team has worked hard and deserve the best we can give.

What is the toughest part of auditing? The answer varies depending on the specific day it is asked. Overall, although I have had the experience to audit many systems in many different industries, generally, organizations in the food and related industries seek compliance with its primary focus on using the structure and discipline to build a quality management system. They are doing this knowing that the structured quality management system will provide the tools to bring its process operation into a top level of performance. These companies for the most part are not focused on obtaining a certificate on the wall but want what the system will do for them. Generally, they are like sponges wanting to absorb as much information as they can to make its system the best that it can be.

Unfortunately, there are those operations that literally must have the certificate on the wall to do business. Dealing with these companies is hard because people just want to get it done and see no value in it. A day of auditing these companies can be filled with many "heated" discussions and angry comments. I always ask the auditee not to "shoot the messenger." As auditors we look for the evidence required by the ISO standard. As emphasized in Chapter 9.5 on Internal Quality Audits, an auditor should have a copy of the ISO standard for reference at all times during the audit. As a third-party auditor, we ask questions based on the statements in the ISO standard and consider how the operation's system has addressed these requirements. I sometimes joke with an auditee explaining that technically speaking as auditors we aren't allowed to think about things. For example, ISO 9001: 1994 states that "incoming material will be verified to ensure that it meets requirements prior to use." It is up to the operation to define how this is done as applicable to its system. If the area work instruction states that a sample is pulled every 5 minutes, then as an auditor, no matter how impractical or useless I might feel the activity, I have to confirm that the sample is pulled every 5 minutes. The area supervisor may get very angry as I inquire and request proof of the activity. If the instruction says every 5 minutes and they are doing it once an hour, this can be a serious nonconformance. It does no good to shoot the messenger or criticize the ISO standard. It was the operation that defined this and as an auditor, I must conclude that it is important to the process because it is defined in the system's documentation. Personally, based on my experience, I might feel that once a day is enough, but as an auditor, I must audit the system as defined. When defining your system, make sure that what you are requiring is really what you want to require. As a consultant, I always warn my clients not to shoot themselves in the foot by overstating the true requirements. I remember once when being audited, we received a noncompliance because we had defined that a particular frozen product had to be maintained at 5°F when in reality it was being stored at −18°F. It was not the auditor's responsibility to interpret for our industry that in fact the latter temperature was much better. It was not as we had defined it. Do not leave requirements up to interpretations because interpretations create variances and variances create potential system weaknesses.

The major downside? The amount of travel required as an auditor is really the major downside of the profession. It does help that the schedule is usually planned as much as 2 to 3 months in advance. In previous professions, there were days when I'd go to work in the morning only to find out I had to be on a plane later that day. Personal plans, birthdays, concert tickets never really came into play. With the lead time for the trips, generally we can do some planning. And although requirements with registrars are different, I was fortunate to be affiliated with one where, when not traveling and performing assignments, I was at home. Although time at home was actually working, making phone calls, preparing paperwork, it was still a great feeling not to have to get up after a long week of traveling and auditing to go into an office.

On the other hand, hardly a day goes by that someone doesn't tell me how lucky I am to get to travel everywhere. I always just smile and think that if they only knew. Generally speaking, traveling involves airports, airplanes, rental cars, maps, the auditee's facility, and the hotel room. Even in an exotic city, I am usually too tired or have too much paperwork to do to really enjoy the city's ambiance. Extra days are usually not in the schedule, and even if they are the travel is so constant and intense that basically the main focus is just to complete the assignment, travel to the airport and make whatever connections are required to get home as soon as I can.

Airplanes and airports do get old. Weather is always an issue. In the winter it is the snowstorms and NorEasters. In the summer it is thunderstorms and the threat of hurricanes. Travel over the years has really become more complicated, but fortunately when one travels this much, it's possible to get the benefits of an airline's upper-level frequent flyer program. At the very least, this gets us on the airplane early or in a reasonable location on the standby list after a canceled flight. Flying every week and dealing with the airlines, is so much fun, that when asked what my least favorite airline is, it is usually the one I flew last week.

A week in the life of an auditor may begin on Sunday trying to travel to the correct time zone for an early morning start on Monday. Although there is some Sunday travel, many times it can be avoided; however, those 6 AM Monday morning flights are many times a too frequent occurrence. Activities when we arrive at the auditee sight vary with the assignment and with our familiarity with the organization and its system. First-time visits can be apprehensive, not knowing the people and the system; but most of the time, the initial visit is a good experience. It is important the auditor presents himself or herself in a professional manner and as being part of the team at the company to be audited. For an initial assessment, we are there to confirm the system has been established, is being controlled, and is maintained in an ISO-compliant manner for the scope of the assessment. During the surveillance audit we confirm that the system is being maintained through internal audits, management reviews, and corrective and preventive action activities. As auditors we are not there to be on a witch-hunt to see how many things we find wrong, but we do have to report the status based on our evaluation. Just because we like the people or perhaps wish we weren't there (it does happen), we must be fair, objective, and professional with our every word and action. There are some days and some audits that are harder than others to accomplish.

In the food industry, it is hard to trust an auditor and truly believe that he or she is there to help. Many different organizations, co-packers, and regulatory agencies may audit a food industry company. Frankly for most of the non quality system audits that I have experienced, it became evident very fast that the auditor was not there to help. As an auditor, I encourage auditees to ask if they don't understand the question. Looking forward to the audit as a positive experience is definitely a new way of thinking. Auditing is like a trip to the dentist. It is good to get my teeth cleaned, but getting them cleaned can be a negative experience. Thankfully though, there are only a few places that I visit where the team looks at me with the same look that I give my dentist. More often then not, management has created the culture for the audits, and the auditees have had practice through their internal audit program. They are proud of their accomplishments and by the time our audit comes around are ready to shine. And for the most part, food industry and food-related industry companies that I have had the pleasure to work with do shine. It makes me proud to play a role in auditing and completing the registrar's forms and reports so that the system can be recommended for certification to the ISO standard.

What is the best and the worst part of auditing? The answer would have to be all the great people I meet and the variety of things that I get to see. I have really made a lot of good friends and am thankful for each and every one of them.

What is the worst part of auditing? Without a doubt, it is having to report the negatives. It is difficult to report a "hold" point, especially during an initial or "main" assessment. Members of the auditee team have usually given such blood, sweat, and tears into its program that I hate to see their disappointment. But as auditors, we must do our best to be consistent and accurate. If an aspect of the system is not compliant, then we must report it as such. Auditing is tiring because we are on our feet most of the day, thinking, asking questions, and being professional at all times. It is difficult when the auditee is very aggressive and wants to argue every point. I do encourage the auditee to challenge statements, ask for explanations, and provide additional information on the issue as useful to the situation. The key is to understand and base the discussions on fact. As an auditor, we do not know the system; we ask a lot of questions searching for confirmation that the system is defined and compliant.

Many times the information is not in the area we are examining but is available and can be presented. This is so important to the total success of the audit. Keep in mind though that "challenges" based on fact are much more effective than constant arguing and aggressiveness. This only wastes valuable time the auditor could be using to follow confirmation trails. Negativity and exercises of verbal Ping Pong wears down the entire team.

In conclusion, what is the worst part of being an auditor? By far, this auditor feels that the worst part of auditing is the travel and having to be away from home for so many nights. Why do people become auditors? For everyone it is different, but meeting great people and being exposed to so many different systems and operations are, I am sure, the top reasons.

12
ISO 9000: THE "ENVELOPE" FOR THE FOOD INDUSTRY (HACCP, GMPS, TQM, MALCOLM BALDRIDGE, AND MORE)

Jon Porter (President, J. Porter and Associates, Ltd.) made the statement that "ISO is the envelope and everything else fits inside it." Don Corlett is quoted as stating that "ISO is a good house for HACCP to live in."

This chapter begins with an article that was written by this author and published in *Food Quality* October 1996. Pages 17–18. It is reprinted with permission from *Food Quality* magazine, ©1996 Carpe Diem Communications. Although this was written in 1996, its focus on the role of ISO 9000 in relation to other food-related programs can still be applied in the year 2000 and beyond.

ISO 9000, HACCP and GMPs: The Family Tie

As the US food industry's interest in the ISO 9000 quality standard increases, discussions of its relationship to existing quality and safety programs, such as HACCP (Hazard Analysis Critical Control Points) and GMPs (Good Manufacturing Practices), are becoming more prevalent. The food industry and relevant government agencies have been developing standards and regulations for many years, and as quality professionals continue to strive for improvement, we must be careful not to reinvent the process and the progress made.

The ISO 9000 standard is one of the more recent programs to enter the fold of quality management in the food industry. Essentially, this standard brings a structured quality management system to the existing family of quality and safety programs, providing a strong, effective and disciplined system that promotes customer satisfaction and continuous improvement in quality. To satisfy my own curiosity and personal interest in the food industry, I usually ask ISO-approved companies during a six-month surveillance visit what they feel have been the benefits of an established ISO 9000 quality management system. The responses given most often include improved efficiency through better documentation and communication, reduced rework, increased customer satisfaction, reduced customer audits, improved motivation and employee involvement through all levels of the process, and improved positive management control.

Since HACCP development and compliance is current in all our minds, let's first briefly look at its relationship with the ISO 9000 standard.

Which Comes First

ISO 9000 and HACCP both focus on preventing, rather than detecting or correcting a program during final inspection. However, an important difference is that HACCP focuses on the product and ISO 9000 on the system. The development of the HACCP plan identifies critical control points and procedures and/or activities identified to adequately control them to ensure safe production of a food product. The ISO 9000 quality management system provides the structure and foundation for the maintenance of the quality system, and, as such, certification to an ISO 9000 conformance standard does not certify the product. It does certify that the approved company has a quality system that meets the scope of the stated standard.

ISO or HACCP first? This commonly asked question depends on your individual situation and circumstances. If you have an ISO quality system in place, then incorporating the HACCP plan into this structure can be achieved quite effectively since the system will already have the required structure and disciplines. Many of the ISO approved companies I have been associated with have done exactly this.

For many, however, the HACCP plan already exists. HACCP is a good precursor to ISO 9000, though it is very important not to reinvent the activity, but rather incorporate it into your emerging ISO system as you would your other process activities. This concept should also be applied to existing programs such as GMPs, pest control, sanitation, and food hygiene, all of which can be identified within the ISO 9000 quality system. This exercise can also be used as an opportunity to evaluate and improve any existing programs in that you can ensure that the activity truly reflects what you want it to do.

During the development of the quality system, typically it is necessary to clearly evaluate and define each activity. Confirmation that performance is being performed is required. How many times have we been involved with "hand me down" procedures and activities? We wonder, "why are we doing it this way?" and the only answer is, "Because we always have." Company management on various occasions have told me that one of the most prevalent internal benefits of the quality system is the identification of "non-essential" activities. The development exercise allows an effective means for defining and addressing exactly what is required.

Integrating Programs

Don't misunderstand the relationship. Many of my colleagues stress that individual quality programs should be separate so as not to overburden the system, causing confusion and ineffectiveness. Certainly, one large program may not be the answer and the exact course of development and implementation depends on your process. If you have an existing and effective GMP program that stands alone, then incorporate it into the ISO system as its own process just as you would any other separate activity. Many processors have GMP, sanitation, food hygiene, pest control and other appropriate activity as part of their existing Quality Assurance program. In these cases, the programs are incorporated into the ISO structure and managed accordingly.

One of the major arguments against ISO 9000 in the food industry is that it is hard to fathom what it can add except more work, constraints and paper. But I have seen first-hand from both the processor's and assessor's point of view that there is no doubt that integration of ISO requirements contributes to improvement through its structure and discipline.

In my days prior to focusing on registration as a registrar, I was involved in the development of an ISO program in two manufacturing facilities and one corporate office function. It took teamwork, with the manufacturing associates all coming together to do their part in identifying and integrating their processes into the ISO format. Once registered, I asked one of the plant managers about his perspective on ISO, and his response was that, "I cannot imagine operating without it."

Three Categories of Clauses

ISO is known for its focus on documentation. The guiding principles are to: Write down what you do, do what you say you do, document what you have done, and audit to confirm compliance. Deviations are addressed in the corrective and preventative action systems. The twenty clauses of ISO 9001 can be divided into three categories: Plan the business, control the process, and maintain the system.

Plan the Business: Management Responsibility (4.1) and the Quality System (4.2) relate to the quality policy and defined objectives to achieve them. These apply to implementing policy, documenting the system and the need to control all activities.

Control the Process: The clauses of the standard that place the greatest emphasis on the controls associated with providing the final product or service are as follows: Contract Review (4.3), Design Control (4.4), Purchasing (4.6), Control of Customer Supplied Product (4.7), Product Identification and Traceability (4.8), Process Control (4.9), Inspection and Testing (4.10), Control of Inspection, Measuring and Test Equipment (4.11), Inspection and Test Status (4.12), Control of Nonconforming Product (4.13), Handling, Storage, Packaging, Preservation and Delivery (4.15), Servicing (4.19), Statistical Techniques (4.20).

Maintain the System: Once the system is implemented, it is necessary to ensure that it continues to operate effectively. The clauses of the standard that are significant to this are: Management Review (4.13), Document and Data Control (4.5), Corrective and Preventative Action (4.14), Control of Quality Records (4.16), Internal Quality Audits (4.17), Training (4.18).

These activities bring ISO 9000 into the family by providing the discipline to monitor the process through the structured internal audit program. Corrective and preventative action requirements not only evaluate existing problems for correction and root cause analysis, but evaluate the processes which may identify and address potential problems. Improving the process through continuous improvement saves money and contributes to the profit of the organization.

I encourage each food processor to review existing programs and become familiar with the ISO standard. Certainly, in the current regulatory climate, these activities should be investigated. Many feel that actually achieving registration and maintaining the certificate through periodic third-party audits is the most beneficial course of action, while others have made improvements from just familiarizing themselves with and applying the ISO structure to specific processes. In many cases, though, the latter have eventually gone all the way and achieved registration. The important point to remember is that ISO is not a replacement for other specific quality tools; but rather it is an effective management tool that, when implemented, its requirements will be integrated into your existing processes and will just make you better at what you already do.

From this information, the critical point to remember is that compliance to the ISO standard is not a replacement for other quality programs, but it is a management tool that when used as part of the total program provides structure and discipline focused on producing quality products and meeting the customer's needs and expectations.

ISO 9000 and HACCP

Food Guidelines (1995) states that "indeed they [ISO 9001:1994 and HACCP] are complimentary and HACCP should be the core part of your quality system" (p. 43). Randy Dougherty (Former President, NSF-ISR) made the statement that a food manufacturing company "cannot have food quality without food safety." NSF-ISR is an accredited ISO 9000 registrar. The company felt so strongly on this issue that it actually developed a program in which food companies could be certified to HACCP 9000. HACCP 9000 confirms compliance to not only ISO 9000 but also to the Codex Alimentarius document for defining, implementing, and maintaining the HACCP plan.

Tom Marchisello (Director of Quality Assurance, Campbell Soup Company) expands on this concept:

> HACCP—The Seven Principles of HACCP are closely integrated with the QMS elements of ISO 9000. I believe that a food company should not be awarded certification without HACCP as part of its ISO 9000 scope. A HACCP program may be effective when it is first implemented, but may not be effectively maintained or improved without the disciplined QMS support from elements such as: corrective and preventive actions, internal auditing, calibration, information control, and training.

Hazard Analysis Critical Control Point

The process of developing a HACCP plan and maintaining a food safety program is extremely complex and cannot be fully addressed in this text. However, based on the experience and expertise of this author and frankly many years of exposure to the industry, it is recommended strongly that every food company understand and apply the principles of HACCP to its process. The following text provides a preliminary synopsis for understanding and applying the basic concept of HACCP.

First it is important to understand specific terms that will be used during this discussion and the requirements of the seven principles of HACCP as defined by the Codex Alimentarius Commission. This commission includes a joint effort by the FAO (Food and Agriculture Organization of the United Nations) and the WHO (World Health Organization) to create a document addressing food hygiene and food safety (HACCP) to be applied worldwide as the standard for establishing and maintaining HACCP plans for all aspects of the food industry (2).

HACCP is activity developed to identify and control potential hazards that are critical to consumer safety. The focus of HACCP is on product safety.

More Thoughts on the Definition of HACCP

> HACCP may be defined as a logical system designed to identify hazards and/or critical situations and to produce a structured plan to control these situations.
> —Dr. Ron Schmidt, Professor, University of Florida

> [HACCP] is a preventative system for assuring the safe production of a food product.
> —Dr. Barbara Blakistone, National Food Processors Institute

> HACCP is a management tool directed to control risk and provide safe, quality products while generating profit.
> —Jon Porter, President, J. Porter and Associates, Ltd.

HACCP-Related Definitions

A "hazard" is defined as

> a biological, chemical or physical agent in, or condition of, food with the potential to cause an adverse health effect. "Hazard analysis" [is the] process of collecting and evaluating information on hazards and conditions leading to their presence to decide which are significant for food safety and therefore should be addressed in the HACCP plan. [A] "Critical Control Point" [CCP] is any point in the chain of food production from raw materials to finished product where the loss of control could result in an unacceptable [or potentially unacceptable] food safety risk. [The] "HACCP plan" [is defined as a] document prepared in accordance with the principles of HACCP to ensure control of hazards, which are significant for food safety in the segment of the food chain under consideration. (2)

Related to the definition of a HACCP plan, it is important to remember the phrases "significant" and the "segment of the food chain under consideration." The term "safe" refers to the processing of food products without contamination from any pathogenic organism or adulteration with harmful chemical or physical material.

An example of a CCP would be pasteurization. "Pasteurization" is defined as the application of heat for a specific amount of time to destroy pathogenic (disease causing) organisms. The lower the temperature the longer the time, and the higher the temperature the shorter the time (1). These time/temperature requirements are based on scientific data and focus on identifying the minimum amount of time needed to kill the most heat-tolerant pathogen known to the particular product. In milk, the time/temperature relationship was first identified for the pathogenic organism responsible for tuberculosis. Later it was thought that *Coxiella burnetii*, the pathogenic organism responsible for Q fever, was more heat stable. Thus the time/temperature relationship was revised. Pasteurization is named for Louis Pasteur, the French scientist who in the early 1860s demonstrated that wine and beer could be preserved by heating above 135°F (57.2°C). In the United States, milk, cheese, egg products, wine, beer, and fruit juices are pasteurized (1).

HACCP Principles

There are seven principles of HACCP that relate to developing a HACCP plan as defined by the Codex Alimentarius Commission.

> Principle 1. Assess hazards associated with growing, harvesting, raw materials and ingredients, processing, manufacturing, distribution, marketing, preparation and consumption of the food (3).

This principle requires the conducting of a hazard analysis of the process. The Codex Alimentarius Commission defines five presteps involved in preparing for the hazard analysis:

1. Assemble the HACCP Team: The HACCP team should include representatives from many different aspects of the process such as receiving, blending, maintenance, management, quality assurance, quality control, and so forth. This approach provides insight from associates that are familiar with many different aspects of the process. During this step the scope of the HACCP plan is defined along with identifying the "segment of the food chain involved" and the type of hazards to be included (2).
2. Describe the Product: "A full description of the product" must be documented (2). This includes the product, its processing requirements, storage temperature, and characteristics. Characteristics include such information that will be necessary for evaluation of the hazards such as

the pH of orange juice. For example, "fresh" juice would be stated as such whereas "chilled" orange juice would include the processing temperatures.
3. Identify Intended Use: This activity is intended to identify the use of the product. A citrus product such as orange juice may be consumed by a wide variety of groups such as children, elderly, and immune-deficient individuals.
4. Construct Flow Diagram: The HACCP team must create a flow diagram of the process for which the HACCP plan will be applied. The definition of a flow diagram is "a systematic representation of the sequence of steps or operations used in the . . . manufacture . . . of [the specific product]" (2).
5. On-site Confirmation of Flow Diagram: The HACCP team should make an on-site evaluation of the flow diagram to confirm that it is complete and accurately identifies all the steps in the process. This is important to add credibility and accuracy to the entire process analysis.

As previously stated, "significant" is a key word in the definition of the HACCP plan. As the hazards are assessed, it is imperative that each hazard be evaluated for its potential for risk (the likelihood that it could occur) and the significance of the occurrence (if it should occur, how serious would be the resulting food safety hazard). It is suggested that the risk/significance be evaluated on a high, medium, and low basis. The team must define the criteria for identifying a step as a hazard. For example, a hazard that has a medium risk of occurring with a high significance for outcome may be the defined criteria. Management must decide what level of risk/significance will be addressed.

The Hazards

All the hazards involved with the specific product should be identified. However, as the actual process flow diagram is prepared, only hazards that can be controlled within the process that is being defined in the scope should be addressed. For example, a producer of raw hamburger meat cannot control the cooking temperature of the end user. This can be identified as a potential hazard with communication to the end-user addressed through other programs; but it would not be a specific CCP because it cannot be controlled within the scope of the HACCP plan.

Remember that hazards should be evaluated for each process. What may be a hazard for one operation may not be "significant" for another operation that manufactures the same product for the same intended use. This may be due to equipment or other process considerations. It is a good idea to evaluate and benchmark (compare) your operation with others; however, it is imperative that the plan developed focuses specifically on the operation or process for which it is being applied.

Principle 2. Determine the Critical Control Points required to control the identified hazards (3).

This principle identifies the critical control points (CCP) and how these must be controlled to produce a safe product. The Codex Alimentarius document recommends the use of a "decision tree" for this analysis. This is included within the actual Codex document.

Principle 3. Establish the critical limits which must be met at each identified Critical Control Point (3).

It is these critical limits that must be met. Exactly what is required to ensure that the product is safe must be established. For example, the required time/temperature relationship to destroy the most heat-stable microorganism that can survive in orange juice would have to be identified in order to successfully pasteurize the product. Pasteurization would then deliver a safe product.

Principle 4. Establish procedures to monitor critical limits (3).

The procedures required to ensure that the critical control points are in fact controlled must be established and implemented. "Monitor" is defined as "the act of conducting a planned sequence of observations or measurements of control parameters to assess whether a CCP is under control." Codex Alimentarius further defines monitoring as a "scheduled measurement or observation of a CCP relative to its critical limits" (2). Monitoring of a CCP is essential to the overall control process. Defined criteria must "control" the hazard. The following definitions apply to the use of the word "control" in the world of HACCP:

- Control (verb) is "to take all necessary actions to ensure and maintain compliance with criteria established in the HACCP plan.
- [Control (noun) is] the state wherein correct procedures are being followed and criteria are being met.
- [Control measure is] any action and activity that can be applied and is essential to prevent or eliminate a food safety hazard or reduce it to an acceptable level" (2).

In identifying the monitoring activities, obviously some situations are better than others. It is recommended that targets be defined as a goal. Drifts from these targets (trends in the process) can be adjusted prior to the CCP going out of control. Continuous monitoring such as a recording chart is most often the means of choice, but for many instances continuous monitoring is not possible. When this is not possible, the monitoring activity must be a scheduled event (2). For example, defining frequency as once per shift would not be considered "scheduled." Scheduled would be every 2 hours (plus or minus 10 minutes) during the shift.

Due to the nature and the amount of "delay" to achieve the results, microbial testing is very seldom used to control a CCP. As a general rule, a physical or a chemical measurement provides the information quicker than waiting for microbiological results.

An important requirement by Codex Alimentarius is that "all records and documents associated with monitoring CCPs must be signed by the person(s) doing the monitoring and by a responsible reviewing official(s) of the company" (2). This applies to each identified CCP and must be done upon completion of the activity.

Principle 5. Establish corrective action to be taken when there is a deviation identified by monitoring of a Critical Control Point (3).

A deviation is defined as a "failure to meet a critical limit" (2). Requirements to be performed if a critical control point has become out of control must be defined. This principle requires that the corrective action be preplanned. Preplanned means that should a deviation occur the required action is identified and understood as a matter of procedure. This action must be clearly defined to provide immediate action to protect against a food safety hazard. Associates responsible for these activities must have the training and the authority to initiate the preplanned corrective action immediately (at the time the deviation occurs) in order to protect the product. Codex Alimentarius further requires that "the actions must ensure that the CCP has been brought under control. Actions taken must also include proper disposition of the affected product. Deviation and product disposition procedures must be documented in the HACCP record keeping [process]" (2).

When discussing the term "corrective action" it is important, actually essential, to understand the difference in its use (meaning). The HACCP plan requires that a preplanned corrective action be identified to address a deviation. This means that what must be done to correct the deviation must be identified before the situation occurs. This provides a quick and effective action to prevent any deviated product from being released. A corrective action in terms of ISO 9000 is a process and/or action to correct a nonconformance. These are identified after the finding is documented, assigned, and the root cause analyzed.

Example of Applying Both Principles 4 and 5

In the dairy industry, it is mandatory that there is a flow diversion valve on an HTST (High Temperature Short Time) that automatically diverts the product while sounding an alarm should the temperature of the HTST unit drop below the "safe" set point. It is also required that this be tested every time the unit is started or every 24 hours. There are many checks to ensure its control; however, there still must be a preplanned corrective action should a malfunction occur that might allow raw milk, which contains the potential for pathogenic organisms, to pass beyond the critical control point.

> Principle 6. Establish procedures for verification that HACCP system is working (3).

This principle requires verification and validation of the CCPs. Verification is defined as "the application of methods, procedures, tests, and other evaluations, in addition to monitoring to determine compliance with the HACCP plan" (2). In other words, are all the defined requirements of the HACCP plan being performed. Validation is defined as the act of "obtaining evidence that the elements of the HACCP plan are effective" (2). In other words, is the plan going to provide for safe product? Are the right activities being performed?

Procedures must exist to verify that the established control for a CCP is functioning properly. As previously described in Principle 5, confirmation that the flow diversion valve is performing as required would be an example of this verification activity. The following must be clearly defined in the HACCP plan for each CCP:

- Frequency of the checks.
- Required record or objective evidence providing confirmation that the checks have been performed.
- Identification of responsibility for performing the checks and documenting the results.
- Defined training criteria for those performing the activities.
- Required criteria for and evidence (records) of the review of deviations and appropriate product dispositions.

Validation would be the confirmation that the time/temperature for the CCP accurately meets data for the destruction of the specific target microorganism.

> Principle 7. Establish effective record keeping systems that document the HACCP plan (3).

Records provide the objective evidence that the critical control points have been controlled according to the required procedures. Without objective evidence (proof), the activity is considered as not having been performed. The Codex Alimentarius document provides the following examples of documentation: "Hazard Analysis, CCP determination, and critical limit determination." Examples of records may be "CCP monitoring activities, deviations and associated corrective actions, and modifications to the HACCP system" (2).

Training and Management Commitment

Although training and management commitment are not specifically addressed in the seven principles, they are mentioned several times throughout the text of the Codex Alimentarius document:

> Management Commitment is necessary for implementation of an effective HACCP system.
>
> The successful application of HACCP requires the full commitment and involvement of management and the work force.
>
> Training of personnel... in HACCP principles and applications, and increasing awareness of consumers are essential elements for the effective implementation of HACCP (2).

Training and management commitment are also paramount to the establishment, control, and maintenance of an effective quality management system compliant to the requirements of the ISO standard.

HACCP Implementation

It is very important that each step in HACCP development be followed. Each process within a process will require its own HACCP plan. For example, a processor that producers and packages fresh juice, chilled juice, and frozen concentrate would have at least three separate HACCP plans. It may be possible to define one plan for the activities that all have in common such as receiving. Care must be made in doing this because what may be a CCP for one process may not be for another. For example, fresh juice may require a CCP in fruit washing and grading to control *Escherichia Coli* whereas the pasteurization step for chilled juice may be the CCP to control that microbial hazard.

The risk/severity relationship is paramount to the hazard analysis. This relationship must be carefully evaluated by the HACCP team. Severity of a problem could be high, but the potential for happening extremely low. For example, *E. coli* contamination in frozen concentrated orange juice has, to the best of my knowledge and literature review, never happened. I would challenge a system that has defined the heat treatment on an evaporator as a CCP since there is no scientific data to indicate that this organism could even survive in this product. As a matter of fact, all existing data indicates that it cannot survive. If there is a concern, then this can be monitored as part of a quality program. There would, technically, be nothing wrong with calling it a CCP, but by doing so this creates many requirements of the system, which just may not be practical.

It is important that management understand what is involved in identifying a CCP. By definition, a CCP states that if this point in the process is not controlled, then it will result in a food safety hazard or at least a potential food safety hazard. Thus, any deviation must be addressed; documentation and records must show actions taken, and so forth. Deviation of a CCP is critical and must be addressed. This is why it is so important to carefully consider the commitment that must be given to each CCP. Remember, realistically, there really isn't a right or wrong program. The program is what the operation defines.

Let's go back to the fluid milk operation example. Pasteurization would be identified as a CCP. Scientific data and food laws explicitly provide evidence that if raw milk is not heated to a specific temperature for a specific time period the likelihood and severity of a potential food safety hazard is high. In other words, it is very likely to occur and the potential results could be devastating. If a deviation occurs (the CCP becomes uncontrolled), then a preplanned corrective action would be initiated to prevent the unacceptable product from becoming a safety hazard. Think about the relationship and the potential seriousness when comparing this to the evaporation process described for orange juice concentrate. For the latter example, if heat treatment is identified as a CCP, then the same controls, sense of urgency, records, and the like must be applied. However, there is virtually no evidence indicating that a hazard is even possible in that type of situation.

Management must review the scientific and technological data available and make sound business decisions. These comments should not be misunderstood to imply that non-CCP process activities are not important. They are very important and should be addressed successfully through quality and other programs without burdening the system with the requirements of a CCP.

Due to several recent occurrences of microcontamination in juice products, pasteurization of chilled juice has been a serious topic of discussion. At the time of this printing, pasteurization of orange juice has not been made a requirement by the regulating agencies. Although the occurrences have been rare, the possibility does occur for *E. Coli* and even Salmonella to be present in single-strength citrus juices. Management must review the risk/severity issues and decide if it makes good business sense to identify the pasteurization of single-strength orange juice as a CCP. Controls may be applied as in the fluid milk operation. Some processors may make the decision that this relationship compared with supporting data does not provide enough basis to control with the intensity required for a CCP but will continue to address it as a quality point. Others may make the decision to apply the same controls and logic as applied in the dairy industry. If it were my decision, I would definitely choose this as a CCP. The processes are designed to provide efficient monitoring and safety and with recent developments, it appears that the risk/severity is increasing as the world of microorganisms adjust to different environments.

As recent as 1990, pasteurization of chilled orange juice was a quality issue not a safety issue. Without adequate heat treatment, juice would spoil faster, but these spoilage organisms would not cause a food safety hazard. Times have changed, many types of bacteria have become more resistant to heat over the years. This is why it is important for the HACCP team to validate the program on a regular basis, which would include monitoring scientific data and other resources for new developments and other changes in the field.

Regarding the identification of a physical hazard, some operations identify a strainer and/or magnet strategically placed in the process as a CCP. The process overall may contain several of these strainers, but it would be the last one in the line that *must* catch the foreign object. In identifying these, be careful to state what the safety hazard would be. Would it be any object? If a twig were found, would it mean that the preplanned corrective action had to be initiated and all product placed on hold? Probably not, this would be an indication that the strainer was working and may result in an investigation to determine how the twigs were getting through other quality checks. However, a hole in the strainer, may be a more critical concern. Thus the critical limit for the CCP may be identified as the "hole in the screen."

There are always "what ifs" that can be discussed; however, this may not serve a constructive purpose. It is very important that each operation evaluate its process using the guidelines defined in the Codex Alimentarius Commission's HACCP requirements (1997). Food safety must be everyone's first concern. Each food processor should have an up-to-date HACCP plan that focuses on food safety. This will be unique for each operation. The identification of a CCP should be documented and maintained to meet the requirements of the seven principles.

It is not the purpose of this text to provide a "generic" program that can be applied to any process. This would not result in an effective program. Each processor must understand the concepts and apply them to his or her processes. In reality, at this time, there are no right or wrong programs. It is the program that is developed for a specific system that will provide confidence that the product is safe while making good practical business sense.

Remember that the program must not be static. It requires periodic (at least semiannual) validation and verification. At a minimum, it is during these reviews that the effectiveness of the program, changes in technology and scientific data, and any other practical considerations are evaluated. In addition, processing changes or other activities could result in a special review of the system. This is why it is so important to empower a team with the knowledge and authority to establish, maintain, and evaluate the food safety program. Remember there is nothing more important than to produce a safe product. A product that each associate would feel safe to feed his or her family. Safety comes first. By applying the HACCP guidelines, it is possible to develop and maintain a program that provides confidence while making good business sense.

HACCP/ISO 9000:Commonalities and Distinctions

I had an opportunity to publish an article titled "HACCP/ISO 9000: Commonalties and Distinctions" in *Dairy, Food and Environmental Sanitation* (pp. 156–161) in March of 1997. Although it was published in 1997, it was actually written a year earlier. As times continue to change, there are some definite revisions that I would make in today's world. Below are quotations from that article comparing application of HACCP with the requirements of ISO 9001:1994. Recent thoughts are contained in brackets.

Management Responsibility

A specific review of the HACCP plan, activities, and noncompliances related to identifying and monitoring CCPs can be included in the management review meeting. This is an excellent avenue for emphasizing top management's commitment to the program. In addition, the HACCP coordinator may have responsibilities similar to the ISO management representative. In my experience with ISO certified companies which have developed a HACCP program, this person is often one and the same.

Quality System

Procedures for creating, identifying, and monitoring activities related to the HACCP plan and CCPs may be incorporated as part of the quality plan and the quality system documentation. According to Porter, "Standard operating procedures (SOP) are beyond all tools, for without the SOP there is a void."

Contract Review

Identification of potential HACCP concerns and confirmation that the organization can meet any HACCP requirements should be determined prior to accepting the contract. For example, if a customer requires a CCP, such as a metal detector placed in a specific point in the process, then this must be agreed upon with the organization prior to accepting a contract to produce the product.

Inspection and Testing

Depending on the HACCP plan and the product being produced, the actual performance of a test and confirmation of a positive result may be identified [and defined with a specific CCP focus].

Calibration

As mentioned . . . [previously], depending on the product (risk assessment) identified in the HACCP plan, the calibration of a specific piece of equipment may be identified [and defined with a specific CCP focus].

Inspection and Test Status

If verification of a test status including confirmation of activity by a qualified inspector is identified [specific to a] . . . critical control point in the HACCP plan, then it would [relate directly to the requirements of ISO 9001:1994 Section 4.12].

Nonconforming Product

Identification of problem products [deviation and subsequent handling of the product] at critical control points, their segregation, and disposition may be addressed relative to [the requirements of ISO 9001:1994 section 4.13].

Corrective/Preventive Actions

Having an effective corrective and preventive action system, which includes monitoring trends and root cause analysis, is an essential part of a HACCP plan. ... The prevention of a problem is as important to a HACCP plan as it is to the ISO quality management system.

Statistical Techniques

Statistical methods may be used to monitor the critical control points although limits will need to be well defined so that they do not [violate] critical product safety [criteria].

Food Guidelines Relationship Diagram

The diagram in Figure 12.1 reprinted from the *Food Guidelines* (1995, p. 44) provides an excellent comparison of how the seven principles of HACCP fit "neatly" with ISO 9001:1994.

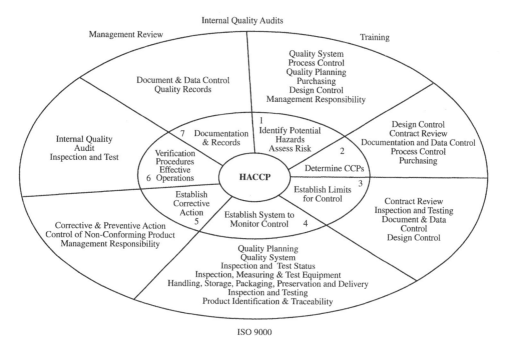

FIGURE 12.1 Using the ISO 9000 Framework to Incorporate a HACCP Plan

ISO 9000 or HACCP

Food industry companies should not try to choose between ISO 9000 and HACCP. Individually they are both excellent programs. Integration of the two can bring the best of both plus much more. There are many similarities and complementary requirements between an ISO quality systeen and a HACCP plan. HACCP focuses on product safety while the ISO standards focus on the overall quality management system. ISO is the total system that provides the structure and discipline to effectively establish, control, and manage all process activities such as a HACCP plan. Let me again emphasize that when establishing an ISO-compliant system integrate the ISO requirements into the existing operation. Although, specific activities may be required, it is not necessary to reengineer an entire process to become ISO compliant. It is necessary to understand, effectively apply, and clearly define how these requirements relate to the process.

HACCP and ISO are fundamental to a process focusing on preventing rather than detecting or correcting a problem. The integration of these valuable "tools" not only makes good common sense but also good business sense. Keep in mind that both ISO and HACCP have a main objective to be proactive, preventing problems rather than fixing those occurrences that have gone wrong.

Thoughts on HACCP/ISO

> My view is that HACCP will compliment an ISO system already in place, meaning that it will be easier to implement HACCP where the ISO system is already implemented. I believe that there should be a more concerted effort to have manufacturers both ISO and HACCP certified.
> —Rex. N. Gadsby, retired GM and CEO Dairy Industries, (Jamaica) Ltd. /Grace Food Processors, Ltd.

As discussed in the *Food Quality* article quoted at the beginning of this chapter, the times are changing. It is time to seriously master the art of doing things right the first time. This can be achieved through effective establishment, control, and maintenance of an ISO-compliant quality management system that is managed and operated by the most qualified, trained associates and includes an effective and well-defined HACCP plan. Although the development of a HACCP plan is still voluntary for some aspects of the food industry as is compliance to the ISO standard voluntary, decisions on whether or not and how to apply these concepts will depend on where the operation sets its critical standards. However, there is no doubt in my mind that to strive to provide the consumer with the highest possible quality food product that is safe, affordable, and makes a profit is the only decision.

ISO First or HACCP First?

The question of which to implement first was discussed in the *Food Quality* article reprinted at the beginning of this chapter. Let us reconsider that discussion and perhaps provide additional insight. Whether to define and implement the ISO-compliant system first or visa versa (HACCP before ISO) is one of the most commonly asked questions. There is no right or wrong answer. Through personal experience, it is felt that having the definition and structure of the ISO quality management system first provides an effective means to develop the HACCP plan. Critical control points and prerequisite programs should be (it is a scary thought to think that they are not) already addressed within the defined requirements for the ISO system. In this instance, a HACCP plan can be developed meeting the requirements of the Codex Alimentarius document with references to the already documented and controlled quality system procedures and work instructions. It is recommended that the HACCP plan be maintained as a separate notebook containing a controlled copy of the related quality system procedures and work instructions.

Other professionals have stated that they felt ISO went smoother because they had implemented HACCP first. The decision belongs to the management of the system, but remember Dougherty's quotation that "a food company cannot have food quality without food safety."

TQM, Malcolm Baldridge, the Deming Prize, and Other Awards of Excellence

Interest in achieving the excellence awards appears to be growing in the food industry. What does ISO compliance have in common with these? Is there a simple answer? Robert Peach (1997) in *ISO 9000 Handbook, Third Edition*, has devoted an entire chapter comparing and applying the strengths and limitations of the ISO-compliant quality management system as related to the disciplines of total quality. Peach states that there must be an understanding of "what the ISO series is and is not." Peach also states:

> Such an understanding will contribute significantly to the correct implementation and use of these standards and guidelines on quality systems and management to meet customer demands for a quality system and to establish a Total Quality system that has direct and long-term beneficial results for the enterprise itself. ISO 9000 is not intended to be a standard for Total Quality. It is, however, a uniform consistent set of system elements and requirements for quality assurance systems and management that can be applied universally within any total quality system. (p. 511)
>
> Comparing the ISO 9000 Series requirements and guidelines, the MBNQA [Malcolm Baldridge National Quality Award] criteria, and Deming-based Total Quality Management (TQM) philosophies and proposed practices is a difficult task. To use a simple analogy, the ISO 9000 compliance standards and the ISO 9004-1 guidelines are like four starched, white business shirts—small, medium, large, and extra large—form-fitting but not expected to cover the whole body. MBNQA is like a giant, one-size-fits-all T-shirt with 24 pockets in which specific

articles are to be placed. Deming-based TQM is like a whole change of wardrobe from which the user is expected to select appropriate apparel for his or her organization. (p. 513)

Keep in mind Peach wrote that comment in 1997 prior to the issuance of even the draft forms of ISO 9001:2000. In the year 2000 as we approach the implementation of ISO 9001:2000, many feel strongly that the standard's redesign and structured increase focus on measuring customer satisfaction. Management commitment and continuous opportunity will bring the similarities with the "excellence" standards much closer overall. Mr. Tom Marchisello [Director of Quality Assurance, Campbell Soup Company] provides an interesting comment on this aspect:

ISO 9001:2000—This revision of the standard appears to be taking quality management systems to the next level. It puts more of an emphasis on system performance and measurable improvement rather than simply compliance. It is taking the basic elements of Baldridge and applies the maintenance discipline throughout the certification and surveillance processes.

ENDNOTES

1. Ensminger, A.H., Ensminger, M.E., and Robinson, J.R.K. *The Concise Encyclopedia of Foods and Nutrition.* CRC Press, Boca Raton, FL, 1995.
2. "Hazard Analysis and Critical Control Point (HACCP) System and Guidelines for its Application." Annex to CAC/RCP 1-1969, Rev. 3 (1997).
3. Pierson, Merle D. and Corlett, Jr., Donald A. *HACCP Principles and Applications* AVI. Van Nostrand Reinhold, New York, 1992.

13
COMMON QUESTIONS, CONCERNS, AND PITFALLS

This text has included specific emphasis on areas of concerns and frequently identified nonconformances. This chapter will summarize some of these areas and enhance the discussion. It will focus on frequently asked questions, areas of concerns, and pitfalls that have been experienced through many on-site evaluations and other inquiries such as those received at seminars and training sessions. Let's explore common questions, concerns and pitfalls!

What Are the Most Difficult Compliance Areas?

Many different surveys conducted agreed that the most difficult areas are document control, purchasing, and calibration.

Controlling documents can be difficult at first just because of the nature of process situations. Typically, memos and other informal means may have been used over the years to communicate information and requirements. Different versions of the same document may remain in many different areas throughout the process. It can be a difficult task during implementation to really get the documents under control. Remember that uncontrolled documents include those copies in filing cabinets, desk drawers, posted, and in associate's pockets.

Bringing the purchasing function into compliance may be difficult because, for many companies, these activities have been performed for some time on an informal basis. The ISO standard requires that suppliers are identified and maintained in a manner that makes sense to the process. It can be even more difficult when many different departments or even corporate offices perform purchasing activities. It is this type of structure or lack of it that enhances the potential benefits for the structure, discipline, and organization provided by the compliant system.

Now let's consider calibration. The toughest part about bringing the system into compliance with calibration activities is identifying which inspection, measuring, and test equipment is actually used to "demonstrate compliance to specified requirements." This can be a long and strenuous exercise, but well worth the effort. Not all inspection, measuring, and test equipment has to be calibrated. That which is used to demonstrate compliance to specified requirements may be classified as "critical." The critical equipment would be maintained to meet all the defined requirements for calibration such as traceability to National Institute of Standards and Technology (NIST) or a known standard; whereas other items, identified as "noncritical," may be maintained to be accurate and operational as process monitoring equipment. These types of decisions relate directly to the nature of the process.

What Is the Average Time from Start to Finish to Achieve Certification?

This question really doesn't have a specific answer. It depends on the nature of the process, what type documentation and system exists, the available resources, and how these are applied. It is very important that a "gap" analysis be performed to determine exactly what exists and the gaps that must be addressed. Management must determine the available resources, which include not only budget and cost issues but also people and time availability. The most important factor when determining the time frame is to make realistic plans based on resource availability and the specific process. One of the most common "mistakes" is to try to do too much in too short of a time period with too few resources. It is best to first identify what must be done and then proceed in a structured, organized, and realistic manner to accomplish this.

What Is the Cost of Certification?

This is also a question without a specific answer. The cost will depend on the existing status of the system, the nature of the business, the size of the organization, and resources available to the process. It can be a very expensive undertaking, but it can also be accomplished in a cost-effective manner through efficient planning and resource allocation. Adequate and effective planning through process evaluations, benchmarking, and training will provide tremendous cost-effective benefits.

Can We Really Expect to See a Return on Our Investment and What Will It Mean to Our Organization?

Mr. Jim Blaha (Plant Manager, Reckitt & Colman) provided an excellent response when this question was asked of him:

We believe that achieving ISO certification was exactly the right thing to do. Being part of the food supply chain, it is extremely important to have a comprehensive state of the art quality system in support of the brand. How can anyone claim to have a dynamic chain without a modern quality plant process? Our key result was in completing the ISO process in support of the brand. Having a great quality process is a key cornerstone in having an excellent brand. From a financial perspective, we have seen a significant reduction in a number of the typical quality measurements: 80% reduction in plant finished goods rejects, 50% reduction in incoming supplier quality rejects, and a 15% reduction in consumer complaints over the past several years. These improvements have provided a down payment on the cost of achieving ISO certification. I believe the future cost savings will significantly exceed the cost of implementation and maintained certification.

How Long Between Actual System Implementation and the Certification or Main Assessment?

This depends on the actual status of the system. Typically, 3 to 6 months of evidence should be available during the main assessment. It is this evidence that the auditors use to confirm that the system has been established and is being controlled and maintained in a compliant manner. Sufficient records confirming compliance must be available for the auditors to review.

What Common Situation Interferes with Companies Passing During the Main Assessment?

The root cause for most "major" or "hold" point nonconformances identified during the main assessment most frequently relates to an immature system. When planning the timetable be sure to provide sufficient time such that system activities and records are available to provide "proof" of a compliant system.

Can the Person with Top-Level Responsibility Also Be the Management Representative?

There is not a specific rule as to who can and cannot be the management representative other than the fact that that person must have "the defined authority for ensuring that [the] ... system is established, implemented and maintained in accordance with [the] ... standard" (ISO 9001:1994 4.1.2.3) Many times, in a small organization, the president or executive manager will also be the management representative. In larger companies this may not be practical.

Do We Need a Preventive Maintenance Program?

ISO 9001:1994 Section 4.9g states that as part of the requirements for controlling the process that suitable maintenance to ensure process capability is performed. Although many systems choose to accomplish this through a defined preventive maintenance program, it is not specifically required. What is necessary will depend on the nature of the business. A product with an extended shelf life and one year's inventory would not have the same requirements as a milk company producing a product with a 12-day shelf life. The latter may not be able to tolerate a shutdown of even a few hours.

What Do We Do If We Do Not Understand the Auditor's Question?

It is very important that the auditee understand the question prior to answering it. Train the auditees to ask the auditor to restate the question. It may also be helpful for the auditee to rephrase the question to the auditor in a manner that confirms an understanding of the question being asked.

Can One Disagree or Argue with the Auditor?

As a general rule, it is not productive to consistently argue with the auditor. This is not only a waste of time but will also create an uneasy or "stressed" situation. It is best to approach a "questionable situation" in a calm and professional manner requesting further clarification including reference to the standard. Be sure when challenging the auditor that the challenge is based on fact and presented in a professional and respectful manner. Many times the situation only requires clarification and further review of records.

What Should We Do If We Do Not Know the Answer to the Auditor's Question?

The best answer is merely an "I do not know." If it is appropriate, then the offer can be made to find out the answer. This does depend on the situation. The auditee should be able to provide information on his or her areas of responsibility. An example of this may deal with records. If a filler operator completes an area form, the auditor may ask how long it is maintained as a record. The auditee may not be responsible for anymore than completing the form and turning it into the supervisor's office. This is what should be discussed with the auditor. It is fine to say, "I don't know." Actually it is the only answer if the auditee truly doesn't know.

What Is the Role of the "Guide"? How Involved Can the Guide Be in the Auditee–Auditor Dialog?

The guide is just that. He or she will guide the auditor to each area of the audit and introduce the auditor to the area representative. Generally, the guide should be available to aid in follow-up and any other type activities that make for an efficient audit. Many times the guide is either the management representative or another member of the ISO team who is very knowledgeable of system activities. Be careful that the guide does not use this knowledge to try to answer questions or otherwise interfere with the auditor and auditee relationship. The auditor will want the auditee to answer the questions. In some instances, too many interjections from the guide may force the auditor to ask the guide to either reframe from answering or step back from the conversation. Most auditors do this in a professional manner, but it can add to distractions and time-keeping problems. Should the guide have comments or guidance, these are best presented on a one-to-one basis with the auditor. Sometimes the auditor may ask for clarity from the guide. The guide should be cognizant of the needs and the direction offered by the auditor.

What Is the Deadline for Upgrading Our Certification to ISO 9001:2000?

Three years from the official date of issuance for ISO 9001:2000. However, it is recommended that management begin implementation activities as soon as possible to provide sufficient time to perform its analysis and collect data that can be applied to the revised system.

What Happens after This Deadline?

After the expiration date, all ISO 9001:1994, ISO 9002:1994, and ISO 9003:1994 will become null and void.

What Would Be the Best Way to Address the Fact That We Do Not Do Design-Type Activities but We Want to Be Certified to ISO 9001:2000?

ISO 9001:2000 provides for "permissible exclusions." It will be necessary for the registrar to review the system and confirm that the permissible exclusions apply, in which case the scope will clearly reflect this situation.

What Can Be Done to Really Get the Interest of All Employees?

It is important to have the employees from all levels of the operation feel a part of this process. This can be accomplished through structured communication programs and hands-on involvement. Let the employees know through whatever means that they are a critical part of this system.

Would It Be Proper to Offer Incentive Programs? What Type of Programs Would Be Appropriate?

Many companies actually create incentive-type programs. Some provide raffle tickets as acknowledgment rewards. For example, a monthly raffle for prizes may be held. Golf shirts as giveaway rewards or prizes are also very popular. Several companies that have on-site cafeterias have given out coupons for free lunches. These are all initiated as a result of associates being involved with system activities.

When Issuing a Corrective Action from the Audit, Would It Not Save Time if the Auditor Just Recorded the Cause of the Problem?

I doubt that this would save time and in actuality may even increase the time. The auditor, being independent of the area being audited, most likely is not in a position to "know the cause." This type of situation may also place the ownership for the fix on the auditor rather than the responsible department associates. The responsible associates are in the best position to identify the root cause and to make a commitment for its fix based on the available resources and the potential risk of the situation.

What Is the Recommended Frequency for Internal Quality Audits?

Although many systems successfully maintain compliance with annual audits, it is recommended that each area and each element (some may be performed in combination) be audited at least once every 6 months. This frequency provides the opportunity for a smaller focused audit compared to the annual audit. It also provides the opportunity to identify existing and potential noncompliances before excessive time elapses. Additionally, it provides an efficient time period between audits to complete corrective actions and to follow up on the actions taken to confirm effectiveness. Should something in one audit be identified that actually interferes with the scope of the audit, revisiting the area in 6 months or so provides the opportunity to evaluate these items.

What Records Should Be Maintained for the Internal Quality Audit Function?

This depends on how the system is defined. At a minimum, it is recommended that the internal quality audit records include the audit schedule and any documentation such as the checklist, observations, summary of activities, corrective actions, and the like that are generated during the audit. It is advantageous to provide as complete a record as possible that may be used as a reference and for guidance for subsequent audits.

Since We Have Always Assigned a Number Grade to Our Audits, Can We Apply the Same Concept to the Internal Quality Audits?

The ISO standard does not state that this type concept can't be used, and technically the system is yours to define. However, it is cautioned that assigning a grade may promote a negative connotation to audit findings. The internal quality audit should be promoted as a positive, non-fault-finding exercise. Findings should be considered "areas of opportunity" based on system activities.

Does the Third-Party Auditor Confirm the System Is Compliant During the Surveillance Audits?

Typically, the auditor confirms the system is compliant during the initial or main assessment. Once this is completed, a certificate is issued and the approved company is identified as being certified or registered to the ISO standard according to the defined scope. For the most part, the surveillance audits review "system maintenance" activities to confirm that the organization is maintaining the system in a compliant manner.

As a Department Manager do I Have Any Responsibilities Other Than Addressing Audit Findings Between the Actual Audits?

It is very important that department management and associates maintain their departments in a compliant state. The role of the internal auditor is to act as an independent set of eyes to assist with compliance. It is not up to the auditor to find everything that is wrong within the department. Department management and associates should continually be cognizant of existing and potential noncompliances addressing these as they surface. Do not wait for the audit to find and report items that you already know must be addressed. Make effective use of the system's corrective and preventive action process.

What Would Be a Good Means to Evaluate Our Process for Meeting Customer Requirements and for Measuring How Well We Are Doing?

This depends on the specific system. It is recommended that the organization create a cross-functional team to evaluate this situation and to make appropriate recommendations. Typically, this information should be summarized and reported at the management review meetings.

We Perform Actions to Correct Situations on a Daily Basis Such as in Relation to GMP Findings. Is It Necessary to Include Every Item in the Formal Corrective Action Process?

Including every single occurrence may overburden the system; however, it is recommended that results from GMP-type audits be reviewed for trends. These trends should be documented in the system's corrective and preventive action process. This will then be assigned to the responsible department managers, evaluated for root cause, and addressed in a timely manner. The evaluation for effectiveness will confirm whether actions taken were effective with results addressed accordingly. An example would be continually finding brooms improperly stored on the floor. The root cause analysis may be identified as the broom hanger not properly located. Once this is corrected, the situation does not reoccur.

We Use Warehouses Not Actually Owned by the Company to Store Our Finished Product. Do We Need to Include Them in Our Scope of Approval?

The most practical and efficient means to address this would be to identify and maintain these warehouses as approved suppliers of services. The services would be that of "warehousing." Remember that all the related "purchasing" requirements must be met, including identifying criteria for being an approved supplier, evaluating performance, and communicating requirements to the service supplier on the purchasing document.

Do We Have to Memorize Our Quality Policy?

It is important for associates at all levels of the operation to understand the quality policy and what it means to the organization. Associates do not have to "memorize" the statement; however, he or she should have a clear understanding of its basic content and his or her role in achieving it. They should also know the location of a controlled copy (i.e., in the area manual, posted in the team room, etc.) for reference purposes.

Our System Is Completely Paperless; However, from Time to Time We Need to Print Documents for Training, Auditing, etc. Would These Be Considered "Controlled" Documents and If So How Would We Control Them?

An effective means would be to include a footer on the printed documents that would identify an expiration date for the printed document. This could be 5 days from the date that it is printed. Another means would be to stamp the printed document uncontrolled; however, then the printing would have to be confined to those that had access to the stamp. Having a "built-in" footer with an expiration date makes the activity more useful and manageable.

Are We Going to Have to Renumber All of Our Documents to Be Approved to ISO 9001:2000?

That will, of course, be an internal decision based on the needs of your specific system; however, it is recommended that a matrix be created that links ISO 9001:2000 elements to the existing documents. It may be most advantageous to rewrite the quality manual in the new format. This will provide a learning experience and a good reference point. The quality manual may then contain the matrix, which in fact may be considered the road map through the existing documentation.

If We Have a Choice Between an Annual or Semiannual Surveillance Audit, Which Should We Choose?

This depends on the specific system; however, it is urged that the frequency be determined by what is best for the system and not what is easiest for the system. Most organizations actually choose the semiannual frequency, feeling that these visits from the external auditor enhance the overall discipline of the system.

Why Won't the Third-Party Auditor Provide Suggestions and Assist Us in Correcting Some of the Nonconformances? Isn't That What We Pay for?

The role of the third-party auditor is to assess the system for compliance. It is important that this person maintain an independent status. Offering suggestions on "fixing" a noncompliance would cross the line to consultancy. An auditor should identify the noncompliance based on defined requirements. Findings and nonconformances should be reported based on fact and not on an opinion as to how the auditor thinks it should be.

Do We Have Any Recourse If We Have a Problem with the Third-Party Auditor or Registrar?

The registrar is a service supplier and should be handled through performance reviews, as are all suppliers. Problems should be addressed through your defined system. A concern with an auditor or any registrar activity should be formally reported depending on the seriousness of the situation and your system's defined means. Just because he or she is your auditor or the registrar is the registrar does not preclude the fact that you are the customer and should be treated with respect, professionalism, and efficiency. Keep in mind that the auditor is trained to evaluate according to the standard and a nonconformance based on the standard is not necessarily grounds for a complaint. If it is felt that the auditor is not basing findings on the standard, being rude and discourteous, or just plain unprofessional, then a complaint to the registrar's office may be appropriate. The initial complaint may be presented verbally via a phone call to the registrar's management. If this does not correct the situation, than a formal complaint in writing may be warranted. This type of action should be followed for all concerns regarding the registrar's performance.

What Is a Common Concern When Evaluating the Quality Policy and Objectives?

It is the identification of a quality policy that is not really measurable. In addition, quality objectives that do not link clearly to the quality policy nor with the organization's strategic plans will most likely result in an ISO-compliant system that remains "stand-alone" from the "real business" part of the company.

Why Is It Necessary to Have Measurable Goals?

There must be a measuring stick in order to sufficiently evaluate how a system is performing. For example, how could the statement "the best possible product will be produced to meet our customer's requirements" be evaluated without "true" measurements such as percent of customer complaints or nonconforming product? An example would be a statement that one wanted to increase his or her stamina in preparation for a triathlon. To truly evaluate progress in achieving this, specific measurable goals such as running time and distance would need to be set and measured.

What Are the Recommended Frequencies for Holding the Management Review Meeting?

I have seen successful systems maintained with the management review meetings held at many different frequencies. However, if it were my system, I would identify it as a minimum of every 6 months. Most successful systems that perform annual management review meetings supplement these activities with additional mini-meetings held in between the official meeting.

We Are a Very Small Company, Which Makes "Independency" of the Auditors Difficult. Is There Any Acceptable Means to Deal with This?

An auditor must be independent of the areas being audited. Evaluate your situation closely and pull from whatever resources are available. Some small companies use auditors from sister locations or other companies in the area. There may be some unrelated "local" companies that are willing to "swap" the service. A consultant may also be a positive choice. In these instances, it is recommended that internal audits continue to be performed by auditors within the system. Then further define that, at a minimum an annual or semiannual audit will be performed by a person external to the system. This may be a complete audit or a partial audit depending on areas requiring the independency and how the process is defined. Remember that this person must meet the system's defined training requirements for an internal auditor and that records must be maintained to confirm this. Another possible option may be to engage the services of a retiree that may be willing to return to perform the audits.

Is It Acceptable to Use Flow Diagrams as Tier Two Documents?

This depends on the system. Many systems effectively "road map" requirements with the use of flow diagrams. The flow diagram actually references the related procedures or work instructions at specific steps in the activity to provide the "rest of the story."

How Do We Determine What Should Be a Corrective or Preventive Action and What Is Just Day-to-Day Routine Activities?

This is a tough question and truly depends on the system itself. Management and the associate team should make these types of decisions based on what is best and most effective for the system. It will take continual evaluations based on trial and error; however, efforts should ensure that process, product, and system existing and potential nonconformances are addressed. Also, train the whole associate team rather than having corrective and preventive actions issued by only one person or one group. Everyone should use this process! As always, effective training is a must! Keep in mind that this will most likely need to be revised as the system matures.

Should We Write a Corrective Action for Every Nonconforming Product Situation?

This will very likely burden the process if every "single" item is documented in the formal system; however, it is recommended that nonconforming product issues be evaluated for trends at defined intervals. These trends should be incorporated into the formal corrective/preventive action program to provide a structured root cause analysis and all subsequent activities required of this defined process.

Our Third-Party Auditor Keeps Stressing That We Apply the Corrective and Preventive Action Process to "Product, Process, and System" Issues. How Do We Know That We Are Doing This?

This basically becomes more of a "way of life" as the system matures; however, in developing and monitoring the program, ensure that it does not get burdened with every singular issue or "one-off" situation. There may be one-off situations that require the formalized program discipline; however, evaluate the situations and trends with a practical eye and address accordingly. Practice and train on the concept of what is a product, process, or system issue and then decide on the most appropriate means to document and address these. One of the most effective systems I have worked with actually had a cross-functional team that met at a minimum of once every 2 weeks to monitor corrective and preventive actions. This was a means to not only identify potential and existing nonconformances, but also to discuss progress, timeliness, and the availability of resources. Results from these meetings were summarized and presented at the management review meeting. This became a very effective and actually time-efficient process.

How Should We Address Preventive Action Type of Activities That Are Ongoing Such as HACCP, Team Training, etc.?

Examples of these could be included in the preventive action procedure or work instruction. Although some systems address these through specific regularly scheduled meetings, the most popular means is to review these activities directly at the management review meetings. Evidence of this discussion should be documented in the minutes. If it is decided to discuss these specifics at a separate meeting, then activities from these meetings should be presented at the management review meetings and documented in its minutes.

Our Corporate Offices Perform All the Purchasing Functions for Our Company; However, Our Third-Party Auditor Wrote a Noncompliance Because We Merely Stated That This Function Was Not Our Responsibility. Why Did This Happen and What Should Be Done to Correct It?

It is the responsibility of the organization to be in compliance with the defined requirements of the ISO standard. Compliance is not accomplished by assigning responsibilities for these requirements to areas outside the scope of the approval. It is essential that the organization actually define its control to ensure compliance. For example, suppliers may be approved and supplied by the customer; however the organization's control is actually confirming that the supplies meet specification through verification performed during receiving and inspection. Procedures and/or work instructions should reflect this.

What Are Some of the Activities That We Should Be Doing to Prepare for Our Triennial Reassessment?

Each department manager is responsible for maintaining his or her area of responsibility in a compliant manner. It is important that area documents are current and that associates are familiar with their use and content. Associates must be trained in responsibilities for related system activities. For example, if quality assurance personnel are responsible for supervising and evaluating activities performed by the pest control service supplier, then these associates should be trained in the requirements for handling "approved suppliers" and have ready access to related work instructions and procedures.

Do We Have to Retrain All of Our Auditors in the New ISO Revision?

Yes, if the internal auditors are going to audit the system for compliance to ISO 9001:2000, then they must be trained in its requirements. Records must be maintained to confirm this.

We Read That We Have to Be Very Careful as to How We Word Requirements. What Would Be Some Examples as to How "Miswording" May Lead to Problems?

One of my favorite examples is the use of words such as "bi-yearly" or "bi-monthly." Most dictionaries clearly define these as being either twice a year (twice a month) or every 2 years (every 2 months). There is a huge difference in these meanings. Also be careful what is defined as a requirement. If a pressure reading must be 20 PSI exactly, then define it as such; but if a variance (20 PSI ± 5 PSI) is allowed, then state that as part of the requirement.

Our System Seemed to Be Doing So Well in the Beginning, But Now 2 Years after Approval, It Seems Harder to Maintain Compliance. What Should We Do About This?

It is very important that the system changes and grows as it matures. A related analogy might be that of a child. The system in its early stages may be compared to a baby who requires focused attention based on the child's age. As the child grows and matures, required attention changes, the nature of the situation relates to the age and maturity of the child. It would not be effective or practical to be applying the same concerns and requirements of a 2-year-old child to that of a 16-year-old. The system, its issues, and problems also change as it matures. It is essential that the management of the system keeps up with its growth and changes as it matures.

Should We Set Up a Specific Time Period to Review All the Documents? What Would Be Sufficient? What Could We Do to Provide the Evidence That This Has Been Completed?

ISO 9001:2000 may require that all documentation is reviewed at a defined period. Prior to this requirement, it just made good business sense to do this. Depending on the system, once per year should be sufficient. Requirements may define that "any document that has not been revised during a calendar year must be reviewed for accuracy by June 30th of the following year." This will provide some flexibility in the process. Evidence of this review could simply be an initial and date on the cover page or first page of the master document by the person with the authority and responsibility to perform the review.

If We Specifically State That a College Diploma Is Required, Do We Need to Maintain a Copy of This for Each Employee

If the requirement is to have a diploma, then evidence must be available to confirm this. In fact a college degree may be the requirement. Evidence of this could be the college transcript rather than an actual copy of the diploma.

Does ISO Make Us Do Statistical Techniques Even If We Do Not Do It Now?

The ISO standard requires that the process be evaluated to determine the need for statistical techniques. For some systems, the use of statistical techniques is just not related to system activities. The standard does not require that the system be reengineered to include statistical techniques as a control mechanism. However, keep in mind that related documentation should clearly identify a defined means (e.g., a minimum of an annual management review meeting) to assess the situation to confirm that its status has not changed.

Can We Hire a Consultant to Write Our Procedures for Us?

A consultant can be hired to perform almost any function within your system; however, please be careful as to what is completed by the consultant. If resources are limited, then a consultant may be beneficial if used to interview associates and guide in the preparation of the procedures and work instructions. Always ensure that the procedures and work instructions are reviewed carefully with responsible associates to confirm their accuracy and completeness prior to issuing.

What Exactly Is Meant by Evidence of "Continuous Improvement"?

This is a term used frequently when referring to processes and systems. Basically, it means just what it says, continuing to improve and identifying opportunities within the processes and system to be "better." In this ever-changing business market, companies must continue to look at ways to improve what they do both in the quality of the process and cost of doing business.

What Might Be a Common Problem in Preparing Documentation During the Implementation Phase?

It is very easy to become overdocumented in a manner that provides excessive detail; however, it is vital that documentation define such aspects as "who is responsible for each step" and "where to record the proof."

What Is One of the Major Concerns Regarding the Corrective/Preventive Action Process?

It is very important that the process not get bogged down by putting every little incidence into the system. This not only burdens the system but also may inhibit finding "the big stuff" that would benefit from the formalized corrective/preventive action process.

Who Should Approve Documents?

The area manager or designee should have some responsibility for related documents. It is up to the system as to exactly which positions it feels should approve documentation. It is cautioned not to overburden the process by requiring too many people for approval. Among other concerns, this may delay the issuance of important system documents.

How Would One Really Sum Up the Overall Usefulness and Benefit of Becoming ISO Certified?

For the most part, I am told from food industry professionals throughout the related field that truly the most useful benefit is the structure and discipline that ISO certification provides to the overall system, which is the basis for continuous system improvements. Jim Blaha (Plant Manager, Reckitt and Colman) provides an excellent summary to this thought:

> One of the greatest benefits that we experienced related to our ISO certification is the significant maturing of the quality system. While we have had several quality improvement initiatives over the past several years, achieving ISO certification provided a comprehensive review and significant advancement of the entire plant quality system from top to bottom. If I were to sum up in one or two phrases what compliance has really meant to our process it would be improved product consistency and doing it right the first time.

We Would Like to Design a Training Program for Our Frontline Supervisors. What Would Be an Effective Means to Approach This Activity?

It is important to develop a program based on the needs and expectations of these associates. It is recommended that the supervisors be surveyed to gather this information and then develop a program accordingly. A sample of the type of concerns that may be included in this survey are as follows:

- List at least three topics regarding the ISO standard and your responsibilities that you would like to know more about.
- At this stage of implementation, what is the status of related documents in your area of responsibility?
- What does the term "quality management system" mean to you?
- List at least five comments (positive and negative) regarding the quality system that you have heard from your associates.
- List at least three positives that you would like to see achieved through ISO 9001 compliance.
- What do you feel has been a negative aspect of your compliance activities?

Please Provide Information on Exactly How an ISO-Compliant System May Fit into Our Process. This Information May Be Used to Enhance Associate Understanding of the Process

The following thoughts may be shared with associates to enhance his or her understanding of the overall system implementation effort.

- ISO reflects worldwide acceptance of a plan for ensuring that standard quality systems are in place in ISO-certified businesses.
- For the most part, certification in food manufacturing companies in the United States and Canada is pursued by those companies that realize that the framework accomplished through ISO 9001 compliance promotes a

structure for growth in overall quality system aspects and in meeting customer's needs and expectations.
- The structure of this type of system promotes continuous improvement and provides a vehicle for positive associate involvement and team building.

What Is ISO 9001 Not?

- ISO 9001 compliance is not a cure-all for all quality problems but provides a system for correcting and preventing problems.
- ISO 9001 is not product oriented. It is not a guide for making better products but merely helps a company document its procedures and monitor activities related to following requirements defined in these procedures.
- ISO 9001 is not a fault-finding, discipline management tool. The focus of ISO is on the system. Nonconformances and corrective and preventive actions are written to identify existing or potential system weaknesses and not to blame individuals. This information focuses on system improvements.
- Products are not ISO certified; the quality management system for producing the products is.
- ISO 9001 is not a complete quality program but provides a structured framework to follow.

Please Provide an Informal Summary of Contributions That May Be Experienced from ISO 9001 Compliance That Can Be Communicated to Our Associates

- One of biggest contributions is that it forces everyone to sit down and take a hard look at procedures and polices.
- It forces consistency so that documented procedures define what is actually done rather than management's "wish list" for the impossible.
- Once a company documents the way "things have always been done," areas for improvement are easier to detect.
- Documenting procedures aids in eliminating the problems that inevitably occur when the person who "keeps everything in his or her head" is unavailable. (I always ask "if you won the lottery and were not here tomorrow, would your replacement be able to perform this function?")
- Documented procedures provide a means to define critical activities within the process, which in turn may be used as a basis for training. This provides consistent training for everyone doing the same job.
- The system requires records to be maintained that demonstrate compliance to defined requirements. This provides evidence to ensure that the

system is working and to provide a means for identifying improvement opportunities and potential problems.

What Types of "Expected Results" May We Communicate to Our Associates?

- Structured system to grow toward our world-class status.
- Structured system to promote team involvement from associates from all levels of the process.
- Consistency in products and in meeting the customer's requirements.
- Where requirements are not realistic, it provides a structured means to address and improve these issues.
- The potential exists for
  - Less rework and product placed on hold.
  - Structured consistency and reliability on equipment through calibration and product testing.
  - Increased efficiency and productivity.
  - Training and problem solving to be easier and more consistent.
  - Problems to be easier to trace due to better record keeping.

Please Provide a Brief Paragraph Describing the Certification Process That Could Be Used to Present an Overview to Our Associates During the Implementation Process

It is recommended that an "introduction" letter containing information as outlined below be prepared and sent to all associates.

> Certification begins by management and associates documenting what is done. Documents must be consistent such that everyone doing the same responsibilities are trained in each document. It is understood that associates doing a particular responsibility knows the job best, that is why everyone will be asked to participate in document creation and review. Once documents are in place, then associates both from your peer groups and from outside will be doing internal audits. Internal audits are designed to confirm the system is in place and to identify improvement opportunities. Everyone will be asked to be part of the team in addressing these items. It is very important that the audit function be viewed as an improvement opportunity and not as a fault-finding exhibition. The auditor and the auditee each play an essential role in implementing and controlling the system.

It is also recommended that a statement of the quality policy and measurable objectives be included. The letter should be signed by top management or the ISO implementation team, which includes top management.

What Are Some of the Most Frequently Asked Questions by the Auditor?

- So what do you do?
- Please explain the process that you are responsible for?
- Are the activities that you have described defined in a controlled procedure or work instructions?
- Please show me these documents.
- How do you know that this is the most current version of the document?
- Do you record any of your activities? If so where is this recorded? Please show me these records?
- Do you use any calibrated equipment? If so how do you know it is calibrated? What would you do if you found indication that this equipment was past due on its calibration?
- Do you ever have instances of nonconforming product? If so how do you address this? Are the requirements defined in a procedure or work instruction? If so, please show me this.
- How were you trained?
- Can you please tell me either as written or in your own words what the quality policy means to you? What role do you play in achieving this?
- Are there any specific customer requirements that you are responsible for?

Is Certification Really Worth the Effort?

Candida Burgos (Process Development Manager, Reckitt & Colman) provides the answer to this question:

> Yes, our quality management system assures that individuals performing specific tasks are fully qualified and are capable of carrying out the jobs and tasks to which they are assigned. Training, process qualification and certification has helped our organization to achieve great results in quality while delivering to the bottom line, cost savings. In the beginning we created a compliant system. Although this quasi "quality system" was in place prior to certification, what we lacked to make the system effective was discipline. Webster's dictionary defines discipline as "training intended to elicit a specified pattern of behavior or character." Certification provided the discipline that allowed us to continually reexamine and apply training in a manner that continually brings out the best in our people, processes, and systems to improve product quality and deliver to the bottom line, cost savings. Our quality system has provided this organization the consistency, direction, focus, and know how to execute the mission (profitability) to achieve desired results.

14
SUMMARY AND CONCLUSION

There is no doubt that there is a strong role for ISO 9001 compliance in the food industry. Experience has shown that adhering to ISO requirements provides the structure and discipline for a quality management system that can be the foundation for continuous improvements, improved quality performance, and increased profits while meeting the customer's needs and expectations. Compliance absolutely makes good business sense. As the world of quality management expands into compliance to ISO 9001:2000, the structure and discipline promoting continuous improvement and customer satisfaction will become even more effective. Never lose sight of the fact that ISO certification does not guarantee quality, but it does provide a structure and discipline that results in a positive framework for the consistent production of a quality product.

> The overall benefit of ISO 9000 is that it provides a framework for the activities which support quality and customer satisfaction. A business can "hang" as much or as little on the different parts of that framework. In the end, all the pieces fit together to form a functioning system designed to deliver quality to the customer. The challenge in developing an ISO 9000 Quality system is to make sure the system created fits the business or process to which it is being applied. If you feel ISO is "forcing" you to do things that don't make sense or to create non-value-added activities, then the standard is not being properly applied. ISO implementation must result in a practical, manageable system.
> —Eric Halvorsen, Quality Assurance Manager Auditing, Campbell Soup Co.

Linda Zastrow, Certified Lead Auditor and current Customer Service Manager, Global Marketing and Sales for Underwriters Laboratory, told me several years ago that "A good auditor can audit any system." Remember this statement as you are defining and implementing your ISO compliant Quality Management System. It is imperative that the ISO requirements are integrated into your system. Do not redesign your system for ISO and especially do not design it for the auditor. It is your system. And just like a tailored suit, the ISO requirements must be integrated into your system.

The Culture Change

I have been asked by many what should be done to promote "culture change." Even though most understand that culture change is required, how to achieve it is the task. It is one of those phrases that puts meaning in the statement that "it is easier said than done." The best book that I have read on this topic is *Leading Change* by John Kotter (1996). Kotter devotes 200 pages to this topic explaining an eight-point process that should be applied in its entirety. Then, discussing "management" vs. "leadership," Kotter states:

> management is a set of processes that can keep a complicated system of people and technology running smoothly. The most important aspects of management include planning, budgeting, organizing, staffing, controlling, and problem solving. Leadership is a set of processes that creates organizations in the first place or adapts them to significantly changing circumstances. Leadership defines what the future should look like, aligns people with that vision, and inspires them to make it happen despite the obstacles. (p. 25)

> Although our plant already has a strong quality system, ISO implementation is helping to fill gaps that tend to be overlooked. Historically, key operators have been relied upon to intuitively run complex systems. By forcing management to closely evaluate and document critical aspects of the process, the system becomes more effectively managed and controlled. Overall, this improves quality and productivity.
> —Brian Dunning, Manager—Quality Systems, The Campbell Soup Company, Sacramento, CA.

System Maintenance

It is very important that once certification is achieved that the entire management team focus aggressively on maintaining the system through management review, corrective actions, preventive actions, and internal quality audits. Successes and efforts must be a result of an entire team focus. Top management must be involved from the start and stay involved. In addition, do not underestimate the role of the internal audit and corrective/preventive action processes within the system:

> The corrective and preventive action element of the system provides a structured channel for action on nonconformities or potential nonconformities. This process

captures the issues, root causes, accountabilities and follow-up that often get lost in informal approaches to solving problems and making improvements.
—Eric Halvorsen, Quality Assurance Manager—Auditing, Campbell Soup Co.

The Team

It is only appropriate that this summary and conclusion restates one of my favorite analogies: A pitcher cannot pitch a no-hitter without the entire team behind him doing the best they can at what they do best. A baseball team doesn't make it from Spring training to the World Series without all members of the team performing their very best through the full schedule of 165 games. Think about the road to the World Series. It includes winning games, home runs, no-hitters; but it also includes losing some games, strike outs, errors, and just some not so great days. To achieve the ultimate, the team played together collectively through all the high and low moments of a season. That is team work. It is also team work that results in a strong and effective quality system that makes everyone proud. Playing in the World Series makes good business sense for the team owners and players, but it goes beyond that, providing the players with a sense of accomplishment and moments to be proud.

> Our success was directly related to dedication and efforts of our team. We believed in our team, we wanted them to be recognized in the global market place. Accomplishing certification brought it all together.
> —Linda Taylor, Training and Quality Manager, Le Meridien Jamaica Pegasus

Bill DuBose, VP of Technical Services, SunPure, LTD stated that the team effort that ISO implementation promotes is very important to every operation. "You do not realize how many different ways one thing is done until the users work together to develop the best single ISO procedure."

Russ Marchiando (Quality Systems Coordinator, Wixon Fontarome) provides us with a great thought to remember for both the great days and the not so great days in the world of quality management systems. This thought was shared in the Preface, which started us on this journey of shared knowledge and experience—a thought that no one in the world of ISO should ever forget:

ISO is a journey not a destination.

REFERENCES

Bolton, A. *Quality Management Systems for the Food Industry.* Chapman & Hall, Gaithersburg, Md. 1999.

Clements, R. B. *Quality Manager's Complete Guide to ISO 9000; 1996 Cumulative Supplement.* Prentice Hall, Englewood, NJ, 1995.

DIS/ISO 9000:2000. *Quality Management Systems—Fundamentals and Vocabulary,* BSI, 2000, London.

DIS/ISO 9001:2000. *Quality Management Systems—Requirements,* BSI, 2000, London. All references in this text to ISO 9001:2000 linked to the DIS version (DRAFT).

Ensminger, A. H., Ensminger, M. E., and Robinson, J. R. K. *The Concise Encyclopedia of Foods & Nutrition.* CRC Press, Boca Raton, FL, 1995.

Food Guidelines (Guidelines for the use of ISO 9001:1994 in the design and manufacture of food and drink.) Lloyds Register Quality Assurance, Croydon, UK, 1995.

"Hazard Analysis and Critical Control Point (HACCP) System and Guidelines for its Application." Annex to CAC/RCP 1-1969, Rev. 3, 1997.

American National Standard Q8402:1993. *Quality Management and Quality Assurance Vocabulary, 2nd Edition.* ASQC. Milwaukee, WI.

American National Standard Q9000–4:1994. *Quality Management and Quality Assurance Standards Part 4: Guide to dependability program management.* ASQC. Milwaukee, WI.

American National Standard Q9001:1994. *Quality Systems Model for Quality Assurance in Design, Development, Production, Installation, and Servicing.* ASQC. Milwaukee, WI.

Kotter, J. P. *Leading Change.* Harvard Press, 1996.

Link, E. *An ISO 9000 Pocket Guide for Every Employee.* Quality Pursuit, Inc, Rochester, NY, 1997.

Newslow, D. L. "HACCP/ISO 9000: Commonalties and Distinctions." *Dairy, Food, and Environmental Sanitation.* Pgs. 156–161, March, 1997.

Peach, R. W. *The ISO 9000 Handbook, Second Edition.* IRWIN Professional Publishing, Fairfax, VA, 1995.

Peach, R. *The ISO 9000 Handbook, Third Edition.* Irwin Professional Publishing, Fairfax, VA, 1997.

Pierson, M. D. and Corlett, D. A. Jr. *HACCP Principles and Applications.* AVI. Van Nostrand Reinhold, New York, 1992.

INDEX

Accreditation, 45, 48, 131, 191, 192
Aldi, Rick, xii, 5, 6, 43, 185, 188
ASQ, 191
Assessment, 25, 47, 48, 49, 55, 194, 217
Atkins, Mark, xii, 10–13, 29

Bay, Rick, xii, 7, 18, 19–20, 23, 40, 188
Benefits, 5–8, 11–13, 19, 21, 23, 24, 48, 49, 58, 66, 72, 74, 95, 110, 111, 131, 163, 165, 177, 180, 186, 189, 198, 216, 229
Bisland, Cameron, 42
Blaha, Jim, xii, 216, 229
Bolton, Andrew, 129, 236
Burgos, Candida, xii, 232
Burness, Mike, xii, 1, 7, 17, 21, 23, 37, 38, 57, 67, 185, 188

Calibration, 41, 51, 75, 104, 109, 113, 118, 138–147, 168, 170, 200, 210, 215, 216, 231, 232
Cartwright, Gail, xii, 8, 36, 38, 57, 186
Castell, Yvette, xii, 7, 20, 24, 25, 36, 42, 185, 189
Certification, 1–8, 10, 15–18, 21, 22, 24, 25, 28, 33–35, 37–51, 57, 58, 66, 67, 71, 81, 86, 95, 124, 131, 144, 148, 151, 163, 165, 180, 184–190, 192, 195, 198, 200, 214, 216, 217, 219, 229, 231–235

Checklist(s), 91–96, 104, 115, 171, 179, 220
Clements, Richard Barrett, 13, 236
Closing meeting, 172, 174, 175
Codex Alimentarius, 75, 200–202, 204–207, 209, 213
Consultant, 10, 12, 13, 34, 39, 42–44, 46, 179, 193, 224, 228
Consulting, 40, 43, 46
Continuous improvement, 3–5, 7, 16, 17, 19, 36, 37, 43, 48, 50, 54, 61, 148, 149, 163, 165, 169, 180, 184, 185, 186, 187, 188, 197, 200, 228, 230, 233
Contract review, 106, 109–112
Corlett, Don, 197, 214, 237
Corrective actions, 9, 21, 71, 84, 95, 115, 148, 149, 152, 157, 160–162, 164, 167, 170, 175–177, 182, 189, 210, 220, 234
Covey, Steven, 11
Critical Control Point, 61, 198, 201–211
Crowley, Dana, xii, 18, 20, 25, 42–43, 58, 74, 95, 112, 160
Customer satisfaction, 4, 5, 17, 18, 31, 37, 62, 87, 122, 131, 134, 196, 198, 214, 233
Customer supplied product, 119–121

Data control, 36, 74, 81, 84, 114, 145, 171
Deming, 182, 213

INDEX

Demone, Dave, xii, 10, 16, 38, 57, 58, 165–166, 186–187, 190
Design control, 60, 108, 127–132
Disappointments, 8, 14, 24, 25, 38, 57, 58, 65, 195
Distribution, 74, 75, 78, 80, 82, 85, 91, 99, 102, 106, 113, 116, 123, 125, 129, 137, 163, 202
Document control, 9, 64, 73–82, 85, 97, 103, 117, 162, 168, 170, 173, 215
Document review, 47, 48
Dougherty, Randy, 200, 213
DuBose, Bill, xii, 235
Dunning, Brian, xii, 234

Effectiveness follow-up, 3, 9, 13, 15–17, 19, 22, 29, 35, 50, 52, 54, 64–73, 78, 80, 84, 87, 123, 125, 130, 135, 148, 150, 152, 154, 158–160, 164, 166, 167, 169, 175, 177, 178, 180, 182, 188, 189, 209, 220, 221
Ensminger, 214, 236
Escherichia Coli, 207, 208
External documents, 64, 75–76, 82, 110, 145, 173

Food Guidelines, 54, 55, 83, 85, 99, 117, 122, 149, 160, 200, 211, 236
Food Quality magazine, 197–200, 212
Forms, 25, 76–77, 111, 152, 158–159
Fowler, Andy, xii, 8, 10, 18, 20, 47, 58, 186, 189

Gadsby, Rex, xii, 7–8, 20, 23, 26, 43, 86, 212
Gap analysis, 13, 46
Garcia, Sylvia, xii, 36, 54, 57, 95, 110–111, 166, 186
Gasser, Keith, xii, 17, 18, 22, 37, 38, 180, 184–185
Gibson, Henry, 7, 17, 20, 25, 58, 81, 185
GMP (Good Manufacturing Practices), 2, 3, 4, 45, 54, 89, 98, 101, 102, 104, 107, 151, 162, 163, 189, 197, 199, 221
Goode, Sue, xii, 4, 6, 17, 23, 25, 36, 186
Gossmann, Al, xii, 9, 15, 58, 116
Gottsacker, Peter, xii, 25, 35
Grandfathering, 94, 95, 97, 114, 115, 117

HACCP (Hazard Analysis Critical Control Point), x, 2–4, 19, 45, 54, 63, 75, 89, 98, 101, 162, 163, 189, 197, 200–214, 225, 236, 237
Halvorsen, Eric, xii, 72, 180, 231, 235
Hazard analysis, 202, 207

Inspection and testing, 21, 41, 51, 60, 75, 76, 99, 103, 105, 113, 119, 121, 122, 127, 130, 132, 135–141, 143, 144, 147, 155, 182, 183, 198, 205, 210, 216, 226, 231
Internal auditor training, 45, 168, 177–179, 181, 191
Internal quality audit, 3, 6, 9, 11, 13, 16, 17, 22, 35, 37, 38, 40, 45, 48, 54, 56, 61, 71, 73, 80, 82, 84, 86, 93, 111, 146, 152, 154, 157, 161, 163, 164–182, 189, 191–195, 199, 200, 220, 221, 224, 226, 231, 234
International Organization for Standardization, 2
IRCA, 192

Job descriptions, 91, 98

Kotter, John, 234, 236

Largey, Dave, xii, 1, 8, 89, 131
Link, Edward, 59, 60, 74, 77, 81–82, 88, 102, 105, 109, 120, 122, 125, 126–127, 132, 143, 146–147, 163–164, 181, 182–183, 236
Lockwood, Bill, xii, 8, 40, 49, 57, 60–61, 67, 72, 126, 131, 148, 185–186, 187

Maintenance, 2, 28, 31, 35, 55, 65, 69, 70, 81, 84–86, 97, 99, 100, 101, 113, 119, 120, 132, 133, 140, 144, 150, 155, 156, 161, 165, 167, 187, 188, 198, 202, 207, 212, 214, 217, 221, 234
Malcolm Baldridge National Quality Award (MBNQA), 28, 63, 185, 197, 212–214
Management commitment, 10, 22, 23, 166, 188, 207, 214
Management representative, 10, 12, 30, 34, 39, 63, 65, 70, 71, 73, 177, 189, 192, 210, 211, 218
Management responsibility, 30, 62–67, 168, 199, 210

Management review, 3, 6, 9, 16, 17, 38, 48, 62, 64, 65, 67, 70, 71–73, 81, 84, 120, 132, 133, 150, 156, 160, 162–164, 166, 167, 169, 180, 183, 185, 187–189, 195, 210, 221, 224–227, 234
Marchiando, Russ, x, xii, 19, 35, 39, 49, 66–67, 68, 71, 96, 139, 148–149, 167, 168, 171, 179, 180, 185, 187–188, 235
Marchisello, Tom, xii, 6–7, 16, 28, 41, 47, 185, 189, 200, 214
Margiotta, Victor, xii, 41, 58, 110, 148, 165
McDonald, Nancy, xii
McGraw-Hill, 4, 191
Modhera, Naresh, xii, 16, 100
Morgart, Karen, xii, 10, 180
Murphy, Jim, xii, 4, 19, 25, 36–37, 49, 74, 81, 131, 152, 184

Newslow, Heather, xiii
Newslow, Martha, xii
NIST, 144, 147, 216
Nonconformances, 21, 53, 61, 73, 82, 86, 93, 97, 101, 106, 107, 111, 115, 117, 119, 120, 122, 125, 126, 127, 132, 133, 135, 137, 138, 142, 143, 147, 148, 151, 152, 154–157, 161, 162, 170, 173, 175, 181, 183, 215, 217, 223, 225
Nonconforming product, 5, 19, 20, 90, 104, 125, 126, 135, 137, 138, 143, 150, 151, 154, 168, 224, 225, 232

Objectives, 2, 3, 6, 59, 62, 63, 65, 68, 69, 71, 73, 87, 134, 171, 173, 189, 199, 223, 231

Peach, Robert, 3, 81, 121, 123, 129, 136, 140, 149, 150, 162, 182, 213, 214, 237
Permissible exclusions, 28, 108, 131, 219
Pierson, Merle, 214, 237
Porter, Jon, ix, 5, 7, 22, 38–39, 40, 50, 187, 197, 201, 210
Preliminary assessment, 14, 38, 46, 47
Preventive actions, 84, 148, 150–152, 160–163, 170, 171, 189, 200, 211, 225, 230, 234
Preventive maintenance, 100, 101, 113, 139–142, 156, 162, 217, 218
Process control, 16, 26, 38, 41, 98–102, 109, 121, 135, 156, 182, 183, 210

Pugliese, Brian, xii, 31–32, 45, 46, 61
Purchasing, 38, 41, 50, 51, 96, 97, 102, 103, 107, 112–118, 120, 146, 170, 215, 222, 226

Quality management, 1, 3, 4, 6, 8, 16, 18, 27, 29–32, 35, 43, 47, 49, 53, 54, 56, 63, 64, 66, 69–71, 73, 74, 87–89, 108, 109, 112, 134, 135, 146, 148, 163–167, 179–181, 185, 193, 197, 198, 207, 211–212, 228, 229, 231, 232, 234
Quality manual, 33, 47, 53, 55, 56, 61, 64, 65, 73, 92, 105, 120, 132, 133, 171, 183, 222
Quality plans, 58–61
Quality policy, 3, 54, 55, 63, 65, 67–69, 71, 73, 88, 89, 98, 159, 171, 173, 176, 199, 222, 223, 231, 232
Quality records, 9, 35, 48, 51, 52, 56, 59, 60, 61, 64, 70, 73, 79–86, 87, 88, 90–103, 105, 106, 111, 113–115, 117, 122–126, 128, 130, 136–147, 149, 150, 155, 159, 164, 166, 168, 173, 176, 177, 179, 181–183, 185, 205–208, 217, 218, 220, 224, 230, 232
Quality system, 1–3, 6, 7, 9–11, 15–18, 22, 23, 29, 33, 34, 36–40, 45, 48–50, 54, 55, 59, 61, 65–71, 73, 74, 77, 78, 81, 83, 84, 88, 95, 96, 110, 140, 149, 163, 166, 167, 168, 172, 173, 180–182, 184–189, 191, 198, 200, 210, 213, 217, 229, 232, 235

RAB, 192
Redditt, xii
Registrar, 9, 18, 25, 40, 42, 44–48, 55, 108, 172, 181, 191, 192, 195, 199, 200, 219, 223
Registration, xii, 3, 4, 9, 13, 15, 27, 32, 43, 45, 49, 74, 186, 199, 200
Root cause, 21, 82, 95, 148, 149, 151, 152, 154–159, 163, 164, 176, 177, 200, 205, 211, 217, 220, 221, 225, 235

Servicing, 60, 98, 132–133
Shipping, 50, 51, 102–107, 110, 111, 121, 128, 138
SIC code, 4, 45
Sladek, Charlotte, xii, 9–10, 44
Smith, Patti, xii, 8, 18–19, 35, 43, 90

Sontagg, Tim, xii, 1–2, 6, 15, 22, 38, 50, 66, 94–95, 140–141, 163, 165, 168, 186, 188, 189–190
Special processes, 99
Statistical techniques, 182–183, 227
Stecher, Charlie, xii, 7, 18, 20, 39, 43, 57, 186
Steven, Ed, xii, 8, 11, 18–19, 35, 43
Storage, 85, 102–107, 110, 113, 119, 120, 121, 128, 129, 133, 161, 202
Supplier approval, 114, 117
Supplier evaluation, 112, 115, 117

Taylor, Linda, xii, 15, 16, 17, 39, 54, 58, 185, 235
Technical Committee 176, 2
Temporary employees, 96, 98
Third party auditor, 171, 172, 191, 193, 221, 223, 225, 226
TQM, 28, 63, 177, 197, 213, 214
Triennial, 47, 48, 84
Training, 9, 11–13, 15, 16, 23, 27, 29, 30, 34, 35, 38, 40, 42, 43, 45, 51, 55, 57, 61, 63, 75, 78, 79, 80, 86, 87, 88–98, 107, 114, 118, 125, 141, 145, 155, 158–160, 162, 166, 168, 170, 174, 177, 179, 181, 186, 187, 189, 192, 200, 205–207, 215, 216, 222–224, 229, 230, 232
Training needs, 88, 92, 98
Training records, 86, 93, 97, 98, 107, 125, 159, 179, 189

Validation, 18, 130, 132, 143, 147, 166, 206, 209
Verification, 60, 104, 117, 119, 120, 128, 130, 132, 136, 210, 211, 212, 226

Weber, Darin, xii, 45–46
West, Jack, 8, 28–29
Wint, Glenmore, xii, 116
Work instruction, 7, 9, 12, 29, 51, 54–57, 60, 61, 72–77, 79–82, 84, 85, 88, 89, 91–99, 101–107, 109–114, 116, 118, 120, 121, 123–128, 132, 135–147, 155, 158, 162, 168–171, 173, 175, 176, 179, 183, 185, 193, 213, 224–226, 228, 232

Young, Ginna, xii, 33–34

Zastrow, Linda, xii, 233–234